IN THE AGE
OF
DISTRACTION

IN THE AGE
OF
DISTRACTION

Joseph R. Urgo

University Press of Mississippi
Jackson

www.upress.state.ms.us

Copyright © 2000 by University Press of Mississippi
All rights reserved
Manufactured in the United States of America

08 07 06 05 04 03 02 01 00   4 3 2 1

Library of Congress Cataloging-in-Publication Data

Urgo, Joseph R.
    In the age of distraction / Joseph R. Urgo.
        p. cm.
    Includes bibliographical references.
    ISBN 1-57806-275-6 (alk. paper)
    1. Attention. I. Title
BF323.D5 U74 2000
153.7'33—dc21   00-021351

British Library Cataloging-in-Publication Data available

CONTENTS

IN THE AGE
OF
DISTRACTION

READ ME FIRST
An Introduction to Distraction

Maybe you should check your E-mail or your voice mail. Is there a program running or downloading on your PC now, something you started before you sat down to read that maybe you ought to monitor? No? What will you do if the phone rings? Something is undoubtedly going on around you now that you may be missing. It's probably just out of earshot or eyesight. Look. Could be that someone is waiting for you to look up or for you to return a call, a message. *The twenty-four-hour headline news has something to tell you.* Surely there must be something else you should attend to now, something to smooth the way tomorrow. Or, for that matter, you are going to be reminded about that soon enough. If you try, you may be able to drive away a lot of these concerns, such as the ones I've mentioned, perhaps. Maybe you can suppress them or sublimate them or else just rest assured that everything is set and in good order and that you may very safely and confidently go on reading. What is needed in this environment is an act of will, if you can locate it, the will to manage distraction.

When I was a child in suburban New England in the 1960s, the routine after dinner was to finish our homework and then watch some television. There was a television room off the kitchen, which we called the den. It was very small and close, and the four of us—Mom, Dad, older brother, me—would sit and watch the situation comedies or dramas until bedtime. On these nights my father would often absent himself after a little while, declaring his intention to go to bed and read. *I'm*

3

going to read in bed for a while. Off he went, to recline alone with one of the two or three books he always had in his nightstand. When my parents were out of the house, I would look at those books. I never saw my father read anything except the newspaper in the public spaces of the house. I've never seen him sit in a chair and read a book. He read the newspaper in the den, in the living room, in the breeze-way—but he never read a book anywhere in the house during the day. To read books he would retire into the bedroom, where he read every night and read two or three books a month, as far as I could tell by the trading he did with people at work. I also never saw my father purchase a book, though he must have. There was an elaborate and informal book exchange in the factory where he worked, a kind of federated library exchange system. The books were mass-market titles, by authors such as Mickey Spillane and Harold Robbins, war stories by Jim Jones. A succession of pulp-fiction texts in heavily read paperback format passed through the house, often grease-stained by hands with jobs that were not clean.

When no one was home I would sit at the nightstand and read through these books in an attempt, I think now, to know my father a little better. He never talked about what he read, and he always read in private (I cannot think of the nature of the emergency or crisis that would have given me leave to knock on the bedroom door when it was closed and he was reading behind it), and so I found the mate-rial that drew him there a tremendous temptation. And then, when I was fourteen or fifteen, he gave me one of the books to read. It was *A Stone for Danny Fisher,* a popular novel by Harold Robbins. From there I read a series of Robbins's books, and we would talk briefly about them, usually in judgmental terms ("that was good"; "this one is better"), and I relished the sense of a shared experience. With my father, speaking was difficult. There was an intellectual gulf not uncommon in that era and aggravated by the invention and perpetuation of youth culture rheto-ric about a generation gap. Although I now recognize the marketing imperatives of such cultural phenomena, they were real enough in suburban households too naive to recognize the sources of such manipulation. We continued to make pur-chases to individuate our interests. But the gesture I shall carry to my last memory is the crimp-covered book with the troubled teenager on the cover. A stone for that character, a book for me. Common reading created a shared sense of distance from a social era that made it difficult to admit to cross-generational longings. The reading assured that each of us had his private life, away from the exigencies of the world and the situation in which we were compelled to live. I knew there was more to each of us than the social cards we were dealt. Civility is born of such mutual recognition.

When my father would take his leave of the den on those evenings, he would always say, "I'm going to read in bed for a while." My mind would immediately and repeatedly stop, and in an important sense, much of the rest of the evening would take place behind a glass plate, separating us from him. The television blared on, my favorite shows anticipated, and my father was going away to read. It meant all too clearly that he had a continuous life elsewhere, a life of the mind inaccessible to us, a private life nurtured and stimulated by a program of reading. It made him more interesting, more connected to my own adolescent desires for rebellion and for escape. Knowledge of my father's intellectual departures added a dimension to him that defied the daily impression I got as he went to the factory to work at his machine, or worked on the cars in the garage, or did household maintenance, gardening, or went to drill with the volunteer fire department. Now, as an adult, I know how important such reading was—and is—to sustaining individuality. Most of what we do with our minds is some variation on the act of following directions and taking orders. Recognition of patterns and fulfillment of expectations occupies nearly all of our conscious efforts while at work or school, at various duties and household chores, and during leisure times. Vacation often means following maps, trails, package deals, or checkout times. Watching television offers a similar set of directives; even if the program lacks commercial interruptions, one is not at liberty to wander away from the set and keep watching. The VCR provides some control over the timing of viewing, but visual stimulation is nonetheless potentially communal and so never truly private. Reading provides refuge and validates the taking of leave from social situations. It insists on privacy and refuses shared distractions; in fact, it refuses to be distracted by social forms. The act of reading imaginative literature thus individuates doubly: once when the reader opts out of the social and again in the act of reading itself, which is an intensely private experience. The intimacy of reading is made plain by the sense of shared experience and private knowledge within the alliances produced by the passing of books.

Now I think I know more about the value articulated by the frequent refrain, "going to bed to read for a while." The real value would not be found in the content of the books themselves. Those who incorporate the serious distraction of reading into the sustenance of their intellects know the relative insignificance of mimetic content or of representation. One does not read imaginative literature for data, to find out things, but for the experience of transcending empirical phenomena. In the middle of William Faulkner's *Go Down, Moses,* a character wants to make a point about the stability of truth in the face of historical and emotional turmoil; to do so, he reads from a poem by John Keats. His interlocutor objects to the selection of the poem: "He's talking about a girl." And the speaker counters, "He had to talk about something" (283). The literary artist always has to talk about something to accomplish the speculative purposes of imaginative writing. Judging by the continuous succession of paperbacks that my father brought in and out of the house, I would venture to say that he had to read about something, too. However, I think now that what he read about had little to do with specific plots and characterizations and more to do with saying "no" to that room and to that house. We were all struggling with the definitions of a mass-market, reproducible family structure architectured into the tightly cornered suburban ranch-style house, struggling with mass-media projections of middle- and working-class existence, and, as the decade drove on, wrestling with the fissures that became wedged into those structures, pressures they could not possibly endure. Reading was an invitation to peace and quiet and escape.

While he was absent, reading, I recall watching *Leave It to Beaver* and *My Three Sons* and wondering where these fathers came from who were so reasonable and articulate and who never seemed disengaged. Seeing these programs in syndication, as an adult, I see one can speculate equally about the children. Models of behavior bombarded our suburban household, right through those episodes of *All in the Family* that had us all, in our well-developed nuclear monad states, seething alone with thoughts about Archie, Michael, Edith, and Gloria. To escape eye contact, I also learned the value of the phrase *going to bed to read for a while.* I remember the first time I said it. It was early in the evening, during the *Ed Sullivan Show,* too early to go to sleep and too late to go out on a Sunday night. There was nowhere to go except Ed Sullivan, nothing to do except to suffer through the distraction of the world of his comparative generational standards of entertainment. Everyone was in the den. I stood up and said it. Generations of boys before me, standing with their fathers over some animal just killed—a deer, a pheasant, a bear—must have had the feeling of entering some equally sacred, human tradition. We weren't hunters. We watched television communally, and we read in private. There was no way out of the room, the house, the era; it was what we had, it was God-given. And so I made my first kill. *I'm going to bed to read for a while.* There was a flicker of recognition in my father's eye, an acknowledgment of something in my mother's, something I would not recognize or comprehend for years to come. The television blared on, Topo Gigio was asking for a kiss goodnight. I think the phone rang. Off I went, to survive in the Age of Distraction.

This book is written out of my own concern that private, internal imaginative space is shrinking as fast as any empirical rain forest or wetland resource. Social reality colonizes more and more of contemporary intellects, clearing space once inhabited by *going to bed to read for a while* and establishing cognitive structures trained to attend to digital cues. The importance of an interior life is denied by technologies that make constant access an imperative, by educational institutions that stress future job preparation, and by media companies that transform news into entertainment. As a result, habits of self-reflection are replaced by continuous acts of social projection. As interior existence gives way to the shared space of Information Age technologies, communities erode and educational institutions continue to evolve into schools of alienation. Twitching students have sat in my office and told me, without self-consciousness, that they do not like to read because it takes too much time, too much thought, both of which are hard to come by, they say. That twitching—eyes darting, looking busy—is a symptom of profound distraction. There's too much to pay attention to, and, even with all efforts marshaled, the sense that something is being missed prevails. But describing this era as an environment has proved difficult because of its pervasive and deceptive qualities. Common sense tells us that what is public, accessible, and real is the stuff out of which the social is constructed. However, experience argues to the contrary, suggesting that interiority, the mystery of what is not spoken but silently recognized and acknowledged, is what forges bonds deeper than social structures and family bonds. To gain access to this interior realm, one must exert control over what is ingested intellectually. Simply stated, one must control the distractions.

Thinking and writing about distraction is like paying attention to reverie. One fear is that attending to it will destroy it, make it go away, like becoming conscious that you are dreaming and then waking up. Some things by their nature can't be thought about without altering them, and thinking about thinking inevitably affects the course of thought. If I am daydreaming, and I begin to think about the nature of my attentiveness or pay critical attention to the content of the reverie, I may change its direction, or I may lose it. Commonly considered, distraction is the absence of attention. But inattentiveness is not the whole of what distraction is. Distraction is also itself a form of attention, a mode of attentiveness more privately conjured or submitted to. When one is distracted, one is still attentive, but one may also be made aware of being attentive to something aside from, or apart from, something else. Distraction is, simply, attention to something other. It is a form of attention understood to exist in relation to a quite specifically recognized lack of, or waning, attention. Distraction is attention to something other than that which, at the moment, is not attended to. Distraction thus depends on something *not thought about* for its existence; it fills the gap left in consciousness when attention to something else is let go. Distraction depends on absence for its existence, as when we refer to a distracted person as absentminded.

Our habits of attention are enforced, trained, and conditioned. There are rules in all social situations about what one ought to attend to; in fact, a social situation consists in large degree of a marshaling of attention. Social engagements are informed by directives indicating how, where, and to what stimuli to direct one's attention. Rituals are created to garnish the attention of groups of people to witness major life passages, to mark milestones, beginnings and endings. Failure to note such events does not mean that they did not occur but that their importance may go unnoticed because of the generally distracted nature of existence. In the workforce, attention is tied to productivity. Corporate meetings are scheduled among workers to assure that attention remains centered foremost on certain projects of importance. Those who complain about "too many meetings" are at once acknowledging the existence of a distracted workforce and chafing under the pressure of attention training. Granting admission of distraction into situations of social attention is, in some instances, a form of rudeness; in corporate settings, yielding to distraction may amount to insubordination. Indeed, some social scientists locate the definition of socialization in attention complicity: the willingness to attend to what one has been trained to attend to rather than something else. There may well be in distraction a vestige of resistance to external controls over what one thinks about and what one does.

To distract is to flirt with the edges of socially acceptable behavior. The implications of the action, *to distract,* are as divided as the state of mind produced by the phenomenon itself. If I am in the process of something important, a distraction may be held accountable for errors or omissions that result—I may curse the ringing telephone or insist in some fashion on my privacy. But if I am in the midst of an unpleasant duty when I am distracted, I may claim to be saved by that same bell, and so I embrace the source of interrupted concentration. Errors and omissions may be, as a result, avoided by a well-timed distraction. Dictionary definitions of the verb *to distract* stress its negative connotations. The *Oxford English Dictionary* refers to phenomena that produce disorder. To distract is to "turn away from a purpose, destination, or position"; to divert from an intention, to stop someone "from attending fully to something." In the most extreme instance, to distract is to throw someone "into a state of mind in which one does not know how to act." In this case, to distract is to perplex, bewilder, and agitate, any of which may be preludes to compromise or disaster. These are singularly antisocial forms of behavior, threatening the integrity and purpose of communal efforts.

Nonetheless, we know that distraction may be sought, actively engaged, and purchased if necessary as relief from ongoing and pressing duties of work, family, and existence generally. If that which I attend to is particularly burdensome, I may seek the distraction of a shopping spree, a movie, a book, or a drink. If one's attentive purpose, destination, or position is undesirable, distraction may, instead of leading one toward dementia, save one from madness. Dictionary definitions affirm the status of some distractions as amusements and sources of entertainment. I think, though, that in an age often called information, when minds are not informed so much as bombarded with data, image, and textual stimulation, our understanding of distraction inevitably undergoes fundamental alteration. As we become more at home with continuous intellectual stimulation, as images, noises, and calls to attention come to be known not as interruptions but as the environment in which we exist, then it becomes necessary to study those stimuli that manage to distract us in the midst of the ecology of din. We may want to harness the ability to choose what will distract and what will remain a matter of environmental white noise, like crickets in southern climates. The concept of distraction implies a hierarchy among the phenomena to which we attend, and as that hierarchy becomes more filled and stimuli compete for places on its scale, a power struggle ensues between the stimuli themselves and the mind that is called on either to arrange its attentiveness or to be at the mercy of the loudest, the latest, or the most felicitous. Minds control very little of their existence; at the very least, they might exert some power at the gates of attention.

Distraction is not a thing in itself but is a kind of barometer and implies a judgment on one's state of mind relative to its environment. The concept of distraction is a device by which we may measure a relationship between the phenomena of attentiveness. Distraction measures a symbiosis, at any given moment, between our minds and the phenomena available to our attentiveness. As I drive down the road, the radio blaring some old song I knew twenty years ago is a welcomed distraction from the tedium of commuting to the office. But when my tires hit a slick spot and my grip on the road becomes tenuous, the radio becomes an unwelcome distraction from controlling the car, and so I turn the radio off. One may also say that as I listened to the radio, driving at first was no distraction from my enjoyment of the music. However, as the conditions of the road worsened, driving distracted me from listening. Which is the distraction, the radio (or the CD player or the cell phone) or the driving? At that stormy moment, the slick road could not be turned off, but the music could, and so the case was easily settled. However, if I were listening to directions on how to handle a skidding car, the sound emitting would not be a distraction but an aid to focusing my thoughts. A value judgment on what holds our attention, then, is embedded in any use of this term. When distraction is unwelcome, one's mind is engaged satisfactorily and with a purpose recognized as desirable and necessary; when distraction is welcomed or sought after, one's mind is not so gainfully employed and may well be engaged in matters that are considered inconsequential and not worthy of one's attention. In such cases, a distraction may save one from a sense of wasted time, like magazines in the waiting room.

Behavioralism is partially a misnomer. Socialization processes instill forms of behavior in us, but more importantly, they instruct us on what we should attend to in any given situation. The school is where one practices focusing attention on figures of authority and aligning one's thought processes with those figures. As one advances, it may become more difficult to locate the figure of authority; it is more likely to exist in some received body of knowledge or tradition. Capitalism directs attention to economies of all sorts, from interest rates to the prices of consumer goods, encouraging minds to naturalize the marketplace and to consider it as a kind of gauge to social and personal well-being. Once attentiveness is directed, distraction is defined as anything that dissents from that direction: whispering in class, looking out the window; being unproductive, or constructing an existence outside systems of consumption. Such activities will be dealt with first at the level of attentiveness. The student may be told to pay attention, and advertising will make every attempt to return one to productivity and consumerism. Only when persuasion fails are more drastic, behavioral methods employed. Thought control lies in the extreme; most socialization requires simple adherence to established modes of inquiry and patterns of attention. Those patterns currently involve multiplicity; and those who cannot divide their attention may find the requirements bewildering. The digital age requires divided attention; surviving it with integrity demands some means of managing distraction and harnessing the cognitive energies it represents.

When we talk about distraction, we often do not know what we are talking about—not because we are stupid but because the state of mind to which the term refers is undergoing great changes. In a universe of silence, before radio and telephone, TV, PC, and the mass media, a distraction, like the sound of the town's first automobile, was a rare thing and thus more likely than not an example of a disturbance to the peace. However, as aural stimuli multiplied and the social environment became increasingly mediated, distractions multiplied concomitantly, and human beings learned to cope, even to enjoy, the divisions introduced into consciousness. Silence, like darkness, is seldom encountered by most human beings unless some effort is made to achieve it or some natural disaster downs power lines. The absence of automobile sounds may make city dwellers anxious. Still, distraction has pejorative implications. We use the phrase "to drive one to distraction" to signal a mind so besieged by unsought stimuli as to lose the ability to choose objects of concentration. To be *distraught* originally meant the state of being distracted to near dementia. But the phenomena to which the term refers are no longer exceptional. Silence is hard to come by. Television is more likely to be a constant rather than an occasional feature of households, waiting rooms, and dining halls. Commercial radio follows us into the department store and on the telephone when we are put on hold; correspondence reaches us on our PC, which once meant personal computer but is now more commonly experienced as a public conveyance. Our lives are wired into stimuli and monitoring devices. What can it mean, in such an environment, to be distracted?

Historians of consciousness speculate that at one time in their development, human beings began to recognize the sounds and voices heard in their heads as their own thought patterns rather than external voices or messages from gods, spirits, or other extrahuman sources. The leap in psychological development represented by this evolution gave birth to individuated consciousness. The sorting out of numerous internal voices into one recognizable pattern of conscious thought set human beings on an evolutionary path of increasingly more sophisticated modes of concentration. At first, the inability to hear a multitude of voices was considered a decline in human stature because these voices were thought to be of divine origin. The inability to hear the gods meant that human beings were stranded, abandoned by spirits no longer interested in human affairs. However, what was experienced as divine renunciation was, in actuality, the beginning of conscious thought, the beginning of a process through which human beings would recognize the source of their authorization in themselves and not in externalities while at the same time becoming increasingly aware of and interested in the external, physical world. Observing scientific progress, Henry Adams noted at the beginning of the twentieth century that "as the mind of man enlarged its range, it enlarged the field of complexity, and must continue to do so, even into chaos, until the reservoirs of sensuous or supersensuous energies are exhausted, or cease to affect him, or until he succumbs to their excess" (487). The mind never ceases to learn its own capabilities and to build on them, to evolve in response to its own discoveries and creations. While the mind enlarges the field of complexity, then, and continues to do so, even into distraction, it must either succumb to an endless, successive, directionless series of engagements or seize control of what stimulates it—not succumbing to the excesses it has created but learning to manipulate and maneuver within an environment of distraction. Individuality, if it exists at all, resides in the extent to which it possesses such cognitive powers of will.

I do not mean to suggest that the increased experience of distraction signals a new age in human evolutionary psychology. However, changes in the way human beings think are seldom welcomed and are usually accompanied by reactionary caution. Nearly every new intellectual stimulus in the past century—automobiles, movies, rock and roll music—was met with fear and loathing in some quarters, particularly by minds formed prior to the appearance of the new thing. I can recall my parents' outraged reaction when they saw me doing my homework with the record player on; I can still recall my grandmother turning off the TV when the phone rang. Throughout the last century minds steadily divided more easily, and the term *distraction* has evolved in its connotative suggestiveness. More importantly, cognizance of distraction now implies some choice, some relative power over one's environment. To exist in the Age of Distraction is to live within an ecology of continuous, often relentless intellectual stimulation. When human beings lived in the wilderness, it was imperative to know things like how to kill wild animals, how to locate north in the woods, and how to build a fire to last all night long. Today these skills are quaint. In the Age of Distraction, one is better off knowing how to distinguish a corporate enticement from an information source, how to conduct consumer or employment research, and how to process the complex language of legal discourse and other forms of obfuscation. In both environments, however, one must know when to welcome and when to filter out distraction. Weekend wilderness romantics armed with cell phones and headsets still need to watch the ground for danger. Distraction measures the quality and purpose of our cognitive powers. Anything can distract; we can even distract ourselves by what we create and by what we make up in our minds to suit our most pressing needs.

Distraction is understood throughout this book as that which we allow to capture our attention and as the relationship between attentiveness and social, historical, and educational matters. It is infused with the spirit of *going to bed to read for a while,* the will to an interior life in defiance of even the most socially expected acts of communal participation. Absent is the judgment that distraction in and of itself is always good or always bad, a strictly positive or pejorative phenomenon. Rather, distraction is known here as the term or the category employed to pass judgment on what occupies our minds. As a result, the word will be used in new contexts (such as a study of the way distraction is structured into U.S. civic traditions) to loosen its assessment connotations and deepen our sense of it as a gauge to the relative value among competing calls to attention. A news alert may distract one from conversing with one's child; after that, the child may become a distraction from fully attending to the news alert. During the Carter presidency, the first daughter, Amy, was photographed reading a book at the dinner table. There was a public outcry about it, because reading at the table is an antisocial application of one's attentiveness, a serious infraction. Social etiquette columnists weighed in against the family's behavior. But the girl was serious about her distractions, and matters that occupied her parents were not to her liking. Neither was she distracted from her book by the reporters covering the family meal. Subsequently, she was told to stop, or at least the press was not allowed to attend to the Carters' dining habits. As a result, the public lost interest in the first daughter's habits of consumption. The media's concern had to do with lines of authority in the Carter household. Children are to pay attention to their parents, especially at the dinner table, the site of intense family ritual. Inherent in the ability to command attention are issues of power; inherent in the tendency toward distraction are issues of resistance. At stake in either case is the status of the imagination. I want to ask how aware we are of the exigencies of distraction and how much knowledge can be brought to its conditions.

Thus, this book is a meditation on distraction: its salience in the contemporary world; its function, historically and politically; its relation to the formation of cognition in educational institutions; and its value as an intellectual resource. Distraction is ultimately linked to rebellious creativity, through its institutionalized form in language, imaginative literature. The humanities and the arts in general have often been accused of being a distraction, and fiction is usually singled out particularly for diverting attention from more important, productive concerns because of the time it demands and the energies it contains. In the Age of Distraction, we need to examine imaginative writing for its ability to do just these things to human minds. Imaginative creations' capacity to distract, to redirect thought, and to bring pleasure merits our attention when so many other stimuli vie for it. We need to explain the function of creative productions in the maintenance of cognitive control over what distracts and empowers our minds. To this end, I pursue the idea of distraction in four distinct contexts: (1) digital and informational; (2) historical and political; (3) educational; and (4) literary, all the while maintaining the idea that distraction needs to be understood and harnessed, not succumbed to. If there is a fear behind the arguments put forward here, it is that human cognitive energies will, as Adams warned, become exhausted or worse, numb to discernment among the intellectual stimuli that attempt to affect it. High exposure to mechanical sources of distraction may decrease human sympathy and erode the capacity for deeply felt experiences, leaving the mind as fragmented and fissure-ridden as the external world seems to be in an era that fashions itself beyond the confines of even the present, in the postmodern realm of what comes next.

"Information, Please: The Distractions of the Digital Environment," maps the territory from which we emerge cognitively and makes the overarching suggestion that the Information Age might better be understood as the Age of Distraction. The contemporary intellectual environment is a world consisting of multitasking, fully wired homes, hundreds of channels, and this situation has created seemingly permanent structures of distraction within private, cognitive spaces. Our minds emerge from within the information phase of mass media and react to intellectual demands for which older, linear modes of thought may be irrelevant. Distraction, within linearity, is known as an error, a mistake, a tangent—something to guard against and suppress. Within multitasking intellectual environments, distraction loses is pejorative connotation and becomes purely descriptive, if not redundant. (Of course there are distractions. What else is there?) The object now is to rethink distraction and rescue it in time to know what it is and how to think about it in the age of multiple and simultaneous attention spans. As this is written, the E-mail program operating "behind" the word processor checks a server and alerts me if mail originating from specific addresses arrives; the telephone answering machine screens calls and blinks if a message is left; the electronic calendar interrupts my work when it's time to go to a meeting. I have worked hard to pay attention, and undivided attention is a commodity that must be guarded diligently. Distraction, if we are to survive cognitively and not be overrun by electronic stimulation, must be arranged with the same care by which we arrange for shelter, clothing, and food.

A historical perspective acts to deprive the present of the tendency to overstate its exceptional qualities. The present, after all, consists exclusively of what it has brought out of its past. "*Don't Tread on Me:* Democracy and Distraction" acts to correct an impression that is often implicit in observations made about the electronic information era—that the situation under study is unprecedented and that we have no groundwork at hand by which to comprehend it on any other than crisis terms. On the contrary, in the U.S. historical and political environment, contemporary struggle with distraction may also be recognized as a manifestation of a national style, a style that may be traced to foundational documents and to the historic indifference of a people consumed by business and profit-making activities. The national style of distraction even has its major historical sign, in the symbolism of the Gadsden flag, with its motto, "Don't tread on me." There a serpent, in pieces (a sign of major distraction), warns against confusing its multitasking body with weakness. Distraction may well be a source of strength and potentiality as long as its various pieces are able to reconstitute connected thought when purposefulness is required. Absent that, attentiveness is a private matter. There is a tradition in the United States of being left to (and constitutionally protected in) one's private distractions. To be left alone, in other words, may be the highest expression of American desire.

Distraction is something for which one can be educated, and doing so may well become among the more difficult challenges to educational institutions in the twenty-first century. "Teaching to Distraction; or, Education for the Hell of It" confronts the issue of how to prepare young minds to navigate among competing and simultaneous demands on their attention. Distraction is often thought to be antithetical to education. But rather than purposefully countering distraction with education, we might better understand education as an effort to fill distraction with content and to empower the distracted mind toward controlling and profiting from its imaginative wanderings. A good portion of education ought therefore to be about distraction; it might be understood as the national distraction project, making minds capable of mastering an environment filled with various calls to attention, each backed by motivations not all of which are equivalent to their stated purposes. Hence the chapter's subtitle, signaling a need to move away from overtly vocational instruction toward a renewed sense that education shapes cognition as well as prepares for employment. Distraction must be simultaneously defended and guarded against. It may not exist objectively; it is more likely a value judgment made against paying attention to *that* when one ought to be paying attention to *this*. As such, it merits scrutiny from a very young age, along with math facts and grammatical structures. Distraction needs to be harnessed, not suppressed. At stake is the imagination, or the way out of intransigent and unsolvable predicaments.

Inevitably, when we poke around into distraction, we hit against the imaginary. "The Public Value of Distraction" sketches the worth of the single form of distraction over which the mind can exert temporal and cognitive autonomy: the reading of imaginative writing. The digital environment and the public world work against the now, shrinking it to minimal intrusiveness in favor of what's next, what's coming. All eyes are on the future (where the profits are) while the present, the now, is minimized, like an icon on the task bar at the bottom of one's cognitive screen. Contrarily, the environment of literary distraction works to expand the now with the fullness of reverie, imagination, and creative distraction. The purpose of literary study is to make room in the mind now—not at the sound of the beep and not when the message arrives, but now, in the reading. The mind escapes into literary worlds for the pleasures of renewal, so that it can read past what it knows, or is expected to know, and migrate to other cognitive structures of knowledge. Literary study might thus be understood as cultivating the art of a prolonged and extensive now, where literary texts address minds not in the name of attending to some purpose but with an invitation to exploring underneath, beyond, and past the present. One clue to the vitality of this phenomenon is the sustained interest in reading literature that engages the mind in the face of so many competing calls to attention in the present: Oprah Winfrey hosting and plugging Toni Morrison; Don Imus on morning radio and TV promoting personal literary preferences. The pundits' admonitions seem lately to be in advance of the literary establishment, where interpretation and meaning continue to edge out discussions of cognitive expansion and pleasure.

"Speculation and Survival in the Age of Distraction" returns to the idea of privacy as a way to begin harnessing the mind's tendency toward distraction. Reading is a private act whose privacy becomes more apparent as sources of imaginative stimuli become increasingly communal and public. The image of person and book captures insularity. There he sits, nothing between him and the printed word. If he concentrates, if the prose is sufficiently engaging, he cannot be distracted. He has chosen his focus and redefined physical and cognitive worlds by his choice. Not even the fly landing on his shoulder disturbs his concentration. We cannot see what engages his mind or hear what he hears. We can't join him, as we could do if he were watching TV or viewing a painting. If we ask what he is reading, we will get interpretation; if we ask him to read to us, we will get his voice. Some authors become terms to define the experience they provide, like Faulknerian, Kafkaesaque, or Catherian. These writers are known by the prose experiences of reading them, and yet no two persons can be certain they have read the same texts—as thousands of pages of critical analysis prove. Every reader reads alone, distracted or engaged, by his or her own lights, the human singular.

Reading imaginative literature is best understood as managing the span and focus of attention by choosing or not choosing social reality. Reading remains the most available exit ramp leading off oppressive or intrusive social situations. The social world is inevitably burdensome to anyone's private life, the life by which we survive the world. Reading is an anomaly, truly, in educational programs, which stress socialization. We tend to stress the social potential of reading by tying its instruction so closely to interpretation and to writing and speaking. How else to test what was read? The secret, though, is that the act of reading will save you from socialization and preserve within the pressure of public life a nucleus called alone, immune from what in the Age of Distraction is called, in various contexts, full disclosure, the right to know, and intersubjectivity. In the Age of Distraction, privacy is suspect. What cannot be seen or heard, and then ignored or attended to, is suspect. The digital age and Windows technology has made us suspicious of anything not readily available to us on our desktop. The Internet is full of inconsequential personal web pages, as many people seem to embrace the absence of a private, inaccessible dimension to their lives. Media psychologists encourage us all to speak our minds, to tell our lovers and parents and children what we really think and feel, replacing the fictitious forms and manners of social existence with a more volatile basis in emotion and personal temperament. The resulting violence (often described, quite accurately, as "senseless") has been devastating at times, as moods are translated into direct action by people taught to express themselves at all costs. In such an environment, the privacy represented by reading may come to be known as an act of subversion, because what is read is not public and is inaccessible to anyone save the person who reads. Beyond the mimetic content of the experience, we seldom describe what happens to us that seems so welcome, when we choose the distraction of finely crafted written language. Some books or genres of books are read repeatedly by people in a lifetime in an act not of recollection but of reimmersion, in acts that are often intensely private.

I have tried, over my adult years, to give my father a book to read that either I have enjoyed myself or I thought he might enjoy. I have seldom succeeded. He has also recommended books to me that I have either tried and failed to read or knew by reputation that I did not want to read. This troubled me at first. But no more. The mistake I made in those efforts and faults was to assume that the content of what was read mattered. The same fundamental error is made by those who misconstrue "canon wars" for educational debate or who associate the installation of habits of reading with mimetic content, choosing reading lists for students based on the representational qualities of the texts. It is true that in the world of social realities, it matters dearly what one says or does: what is done sometimes matters much more than how it is done. Contaminated water is useless and dangerous no matter how it has been ruined. The reverse is true of imaginative stimuli, where how a thing is made and its effect on the reader's mind is more important than its content. Relevance to the social world of the reader may be the death of imaginative experiences, which, by the definition at work here, exist as alternatives to socialization. Interiority requires continuous sustenance, and those whose interior lives are constructed by chosen distractions, by *going to bed to read for a while,* know that the content of the distraction is insignificant save as the hook by which one reels in one's own mind, liberated for the moment or the hour from its thralldom to information and the world of external stimuli. My father and I now live in different households in different geographical regions, and we require distinct reading matter to maintain our senses of self. Still, there is that recognition that what is here, commanding attention, is only part of the story and that the self possesses an interiority not completely at the mercy of the social, familial, and corporate.

The Age of Distraction is a marvelous one, filled indeed with technological wonders, world-shrinking capacities, and opportunities to see and do more in a lifetime than was ever afforded the mass of human beings. But like previous ages, the Age of Distraction has its deadening qualities, its inquisitions and its sacrifices. By definition, an *information* age privileges external stimuli. One must tread with as much care now as ever before, saving oneself from becoming what the age consumes and what it destroys in the name of its own perpetuation. *He had to talk about something,* and he had to read about something. Only in the intimacy of family relations may one decide to go to bed and read for a while without rudeness or implicit rejection of the group's chosen activity. Walking away from the evening and the age requires an act of will, but it is a willpower fueled by the need for interior reflection and the sound of the self thinking, on and off the page. Communities are forged by external commonalities, but they are maintained as well by the recognition that interior lives run like currents below the realm of articulation; to protect these passageways, cognitive space must be reserved and maintained for self-preservation in an era that possesses the capacity to convince us that the interior self no longer exits. Negotiating such terms within the Age of Distraction is the chief concern of the pages that follow.

INFORMATION, PLEASE

The Distractions of the Digital Environment

America in the twentieth century systematically attenuated its conception of the present to expand and protect the vitality of its future. It did so through electronics, by producing an environment that has become increasingly characterized by informational distractions, starting with the telephone and the radio and continuing with the television, the personal computer, the video game, and most recently, the Internet. Each of these technologies has acted to either trivialize the present by transforming it into a staging ground for subsequent activity or to effectively destroy the present by introducing repetitive, time-consuming entertainment activities as pastimes. The average middle-class household is fully plugged into electronic distractions so that beeps, buzzers, and commercial sound tracks provide community-building aural atmospherics throughout neighborhoods and across the country. What holds communities together is, in part, this shared sense of constant distraction. Watches beep on the hour, beepers request callbacks, microwave ovens make electronic announcements, and the television is always on—such is a portrait of the kitchen, anyway. The sound that is heard the least is silence, and the dimension of time that is most consumed is the present. The future expands continuously. Electronic media constantly tell us what we should be doing instead: answer the phone, be reminded of the time, attend to the newscast, listen to the commercial calls to invest for the long term, plan your vacation and your retirement. Cognitive control over one's now is thus consistently surrendered to electronic arguments for subsequent, not present, action. Americans, who have always had a love affair with the future, have welcomed these demands. *What are you going to do next?* Everybody knows you're only as good as your next move.

The dominant mode of electronic distraction is informational, whether by voice, aural and visual data, image, or text. Morning news, daily telephone calls, voice mail, electronic banking, evening television rituals, E-mail, and web surfing are the common structures that input a plethora of information and data into the cognitive structures of American minds. The age of electronic information has been accepted on false grounds, however, if we believe that we have increased our access to significant information as a result or that we have saved time by adding devices like E-mail or voice mail to our information-gathering routines. Little of consequential value is transmitted by these new structures, and saving time is almost never the result of quickly gotten information. Ironically, despite technology designed to make lives more efficient and to add measures of convenience to daily obligations, most middle-class Americans claim to be busier and to believe they have less time than previous generations as a result of "time-saving" devices. The situation is not surprising. Technology does not save time but defers it, robs from the present, the now, to expand the sense of the future. All things become possible—but always and only in time. Rather than allowing the cognitive *now* to attend to contemplation about its state of being in the present (thus expanding the now with imaginative content), electronic media fill the now with opportunities regarding what to do next—answer this, watch that, click here, or, in the words of the mid-1990s Microsoft slogan, "Where do you want to go today?"

The data content of such informational input is virtually irrelevant, and new electronic technologies repeatedly confirm Marshall McLuhan's observation that with electronic media, the medium, not the information, is the message. Or, rather, the medium constructs and maintains its own content by providing a specific kind of informational experience. It is not the particular web page but the Internet that one desires; the experience of clicking and the pleasurable simulacrum of cerebral travel far outweigh the satisfaction of getting the headlines from MSNBC or making a purchase at a commercial site. We know that people are far more likely to watch television because they want to watch television—not because they want to watch a particular program or gather some particular insight and then switch off the set. Likewise, people are more inclined to have the desire to surf the Internet rather than to log on to visit one site and then log off. The Internet is designed as an informational experience, combining visual, aural, and textual sources. When the telephone was the primary technological device in the home, one could call various recorded messages (such as the weather, the news, or the time) when one had the desire to simply use the phone for its information-gathering essence. Today, the telephone is being gradually usurped for such satisfactions, and even crank calls are giving way to E-mail spamming. But whether one is watching television, surfing the web, or even listening to the radio, the technological content of the satisfaction is equivalent. Each of these media satisfies the desire to acquire inconsequential information. The inconsequential nature of the information is what qualifies the medium as entertainment, because the experience is one of gathering disposable data, not of being transformed by insight or reflection. One seldom passes the time gathering self-destructive or other forms of information that require or demand action. The twentieth-century electronic citizen was proven to crave the experience of knowing, and the cheapest, easiest, and least taxing means of doing so comes via modem, cable, and power lines. The content of electronic distraction is not data in particular but the experience of information itself.

The introduction of various electronic media into household routines preceded their role as information sources. Television was not invented to remedy an absence of news programming. Rather, such programming followed the development of television and created a market need for itself among television watchers who, it was found, enjoyed knowing current events visually. An expressed need for access to instantaneous text messaging in middle-class households did not precede electronic mail; rather, E-mail software, packaged with Internet browsers, successfully instilled a desire for recreational text communication. We would be hard-pressed to argue that any of these media developed in response to problems or deficiencies. Instead, technological developments have contributed to an evolution of distractions, filling leisure time with time-consuming and time-obliterating sources of attention. The telephone answering machine provides a convenient example. The message deposited on the machine shifts the obligation of communication from sender to receiver, so that while the sender has not reached the desired person, the caller has reached an extension of that person in whom, through the technology, an obligation to receive information has been placed. However, in common parlance, one does not claim to have received specific content from the machine but something referred to as message. "I got your message" thus substitutes for a specified, substantive response. One may then say, "and I will answer," to further defer content. The answering machine, which claims to facilitate information exchange, creates an additional layer of formal, mannered, meta-communication between sender and receiver, attenuating the centrality of the present by deferring the conclusion of the exchange into some point in the future, after the messages have been returned and reconfirmed. In addition to occasionally saving time for senders, the answering machine contributes to the removal of telephone exchange from present time to deferred time, thus attenuating the usefulness of the now to a point in which future exchange is planned but seldom accomplished. And checking messages joins the host of distractions preying on minds that would otherwise manage to control their attentiveness.

The chief characteristic of information and communication opportunities in the present is that they are ubiquitous. The television is available twenty-four hours a day, and the Internet expands this access by making all specific program sites available at all hours, overcoming television's reliance on time-specific programming. The two media will eventually merge. The constancy of information distinguishes the present era. The ceaseless quality of information today far outweighs its substance in nearly all but the most dramatically eventful of cases. The incessant nature of television, for example, assures that any particular content contained within its programming will be subsumed by the medium's own continuous format. When information is always available and continuously produced, it is not so easy to select significance, especially when the experience itself is one of being relentlessly informed. An illusion of information is not qualitatively different from the real thing, and this illusion characterizes the information age more than does any particular kind of information produced or disseminated within recent memory. Television critic Bill McKibbon suggests that "if something exceptional happens it hardly matters—it is quickly forgotten, averaged out, eroded by this ceaseless flood." At issue is not the length of TV segments or how much depth or detail is provided. Rather, "It's that each line of thought is instantly replaced by another" (214), and all items are thus flattened out by the onslaught of information. If one were to watch one program or news segment and then turn off the set for a while, it might be possible to isolate and thus fully integrate that information into one's conscious life. Similarly, if one were to log on to the Internet, visit one site for whatever information or activity was desired, and then log off, such access might well be a time-saving convenience, leaving more time for *now*-expanding activities, like reading imaginative literature. However, very few people watch television or use the Internet that way; hyperlinks and television programming work hard to prevent such controlled-use action.

In one of the most influential and damning studies of television written since the invention of that medium, Neil Postman, in *Amusing Ourselves to Death,* bemoaned the replacement of written texts by TV. Postman's argument has become a standard one for educators and literacy advocates generally, and his critique of the medium retains its relevance in the digital era. The insertion of television into American households at the cost of reading, according to Postman, "has dramatically and irreversibly shifted the content and meaning of public discourse, since two media so vastly different cannot accommodate the same ideas. As the influence of print wanes, the content of politics, religion, education, and anything else that comprises public business must change and be recast in terms that are most suitable to television" (8). Television's terms are the terms of mass entertainment, and as a result, "the seriousness, clarity and, above all, value of public discourse dangerously declines" as TV proliferates into the primary medium of public information (29). Specifically, information on television is presented as amusement, designed to condition viewers for consumption, not political action, and as such has emerged as "the last refuge . . . of a culture overwhelmed by irrelevance, incoherence, and impotence" (76). Finally, television production techniques are designed "to encourage us to watch continuously. But what we watch is a medium which presents information in a form that renders it simplistic, nonsubstantive, nonhistorical and noncontextual; that is to say, information packaged as entertainment. In America, we are never denied the opportunity to amuse ourselves" (141).

Postman's claims are not likely to be refuted. Nonetheless, as television evolves and becomes part of a greater electronic network of information sources, including the Internet, Direct TV, cable access, and Web TV, it ceases to be an isolated intrusion into the lives of Americans and emerges as part of an electronic ecology. This ecology, this information environment, must now be navigated by minds aware of the various effects such media have on cognitive capacities. Reading books may have very little to do with this ecology, and threats to literacy represented by information technologies may not differ greatly from past threats, such as labor-intensive and mind-numbing workplaces, seasonal agricultural employments, or the inability to sit still for long. The issue to contemplate is not reading versus television or Internet but a more thorough understanding of the intellectual implications of consuming information for entertainment purposes and of the rapidly shrinking sense of the present produced by such experiences. While it is as crucial as ever to the future of democratic society to encourage people to read and to distinguish the thought processes demanded by reading from those necessary for watching TV, equally important are analyses of the information environment itself, the environment within which reading exists today. Information technologies have not replaced reading; they have moved into the time spaces formerly occupied by the labor-intensive activities most people used to engage in when they worked, shopped, and managed their lives. Today, we spend the greater part of our conscious lives processing information from electronic sources and deferring present concerns to future destinations; within that environment our imaginative capacities struggle, as they always have struggled, for attention.

Before E-mail, Internet, and word processing, there were typed (and retyped) letters, afternoons at the bank, carbon copies, keypunch cards, and a endless series of manual operations at libraries, in offices, and on assembly lines. Before the spread of commonly accessed information technologies, the environment within which daily life transpired consisted of a higher percentage of memory, anticipation, and conversation. In the fifteen-minute wait to deposit a check and make a withdrawal at the bank, the bank patron spent time in recollection, in mapping out subsequent errands, and in conversation with neighbors in line with similar concerns. The time involved could expand well beyond the brief duration as if filled with contemplative experience. One expected the bank stop to take some time, and so the passing of present time in the process was not resented. Today, a two-minute stop at an ATM machine leaves little time for any such activities, and as a result, everyone involved shares a sense of haste, and God help the ATM patron who takes more than a minute or two at the machine. Far from making banking a convenience, the ATM machine has transformed the simple procedures of deposit and withdrawal into annoyances, actions for which patrons hold deep resentment. Anything that takes so little time is an activity of very little value.

While it is possible to save money, old magazines, or stamps, it is impossible and even self-destructive to save time. Time is what invests human activities with meaning. The importance of child rearing is evidenced by the number of years it takes to bring a child to adulthood. The growing complexity of the adult world over the past two hundred years has extended childhood and adolescence (actually, it has invented adolescence), so that what once took fourteen to eighteen years now may consume as much as twenty-five years. Conversely, the unimportance of cleaning infant waste has led to a series of technologies—washing machines, throwaway diapers, wet wipes—designed to allow parents to spend less time on a specific task of little value. Today, changing the baby's diaper is a trivial matter, accomplished even in public places. When time is saved regarding anything human beings do, the value and significance of the activity declines. Before indoor plumbing, a bath was a significant event in human affairs and might consume an entire evening for a family. Contemporary bathrooms make the event an incidental, daily routine, hardly worth the label of ritual. Male shaving once commanded elaborate devices and creams and even occasioned a trip to the barbershop. Today, most barbers refuse the menial task. Menstruation was once a matter of social ritual; today it holds no public significance—again, because of technologies designed to save time. In none of these hygienic examples has the public significance of the activity survived its technological solution, and in no case has the time saved been applied to increasing the value of actions performed. On the contrary, in every case the technology has adjusted the value of the act downward in direct proportion to the decrease in time allotted to it.

When a new technology such as the ATM machine offers to save time with simple banking processes, few people use the machinery as a convenience. To do so would be to do one's banking and then invest the saved time into loitering at the door or extending conversation with fellow patrons. I remember my father spending hours standing outside watering his garden, distinctly aware of what husbandry demanded and of the relationship between water and life. Sprinkler systems eliminate such awareness, making the activity necessary to it seem like purposelessness. While convenient, such technologies lower the rate of time considered acceptable for the tasks into which they intervene. If retyping a business letter to insert a new sentence was understood as a thirty-minute procedure before word processing technologies, today the allotted time for such a task is about thirty seconds. The time saved has not gone into secretarial vacations. On the contrary, because time is a relative concept in human consciousness, a series of adjustments have followed the convenience of word processing in which time saving is a cause, not a result, of consequences. First, the value of a business letter has shrunk in proportion to the time consumed in preparing it. In the age of the typewriter, a letter was an event; in the age of word processing, a letter is an occasion for suspicion. Is it a form letter? Am I being sued? Second, the value of the person who prepares the business letter is diminished. Any task that takes so little time, for which templates, spelling and grammar checks, and even voice dictation programs exist, cannot accrue value to those whose sole job is to complete it. Third, the sense of time associated with letter writing is drastically adjusted, so that anyone who must spend a great deal of time in preparation, composition, or revision of a letter is judged mentally deficient in some profound way. And so in addition to offering the letter-writer a convenience, word-processing technologies rob the letter of its primary source of value, the time it once took to create it, and, consequent to the theft of time, the writer's value is diminished. Today, letter writing is an indulgence, a hobby, like crochet or a long bath, something only those with nothing to do or whose minds are unoccupied may afford.

Convenience is no simple matter. The automobile, once an intrusion into the culture of horse-driven transportation, is now a necessity, like shelter, for life in the United States anywhere outside the most densely populated urban centers. In *The Value of Convenience: A Genealogy of Technical Culture,* Thomas Tierney acknowledges the degree of freedom offered by the automobile and goes on to analyze the even greater degree of needs developed by the cultural ecology of privately owned transportation. Tierney draws a comparison between the development of the twentieth-century automobile and the introduction of homesteading in the nineteenth century. The cultural environment that "developed around the automobile set people in motion and required them to move," according to Tierney, and today "this form of transportation technology binds people to the economic structures of modernity in a manner similar to that which bound the nineteenth-century homesteader." Each example, private homesteading and the ownership of personal transportation technology, "binds the consumer with a debt that must be paid over time" and thus introduces long-term indebtedness into the menu of acceptable household financial obligations. Equally important, the time purportedly saved by owning an automobile creates new, time-consuming necessities in the form of "conveniences" that previous technological environments did not need and could not accommodate. The automobile made long-term credit arrangements a universally accepted condition while creating a multitude of what Tierney calls "possibilities for the consumption of convenience." Automobile culture has produced "fast-food restaurants, supermarkets, drive-through windows of all sorts, from liquor stores to banks," all of which could not exist in a world without cars (Tierney 110). The automobile has thus normalized such objective absurdities as traveling twenty miles to go to work or five miles to purchase a loaf of bread. Either of these trips, which would have been unthinkable with horse travel, is considered reasonable and mundane by automobile. And the bread shop, open twenty-four hours a day, is even called a convenience store.

Instructive parallels may be observed between transportation and information technologies. Americans in the past have gathered enough information to confront depression, civil war and abolition, territorial expansion, even revolution without the aid of television news broadcasts, web sites, or CNN. Rarely do we come across historical situations where the absence of constant information interfered with the ability of citizens to make decisions or take political actions. There is no evidence that the average citizen in 1929 or 1829 was less informed than the average citizen in 2000—the truth may be to the contrary. There is evidence to suggest that reliance on broadcast news for information reduces one's knowledge about contemporary events by confusing message and medium. As knowledge of media manipulation becomes more common, so does suspicion regarding the credibility of information sources. Mass entertainment is now made from plots involving phony information, such as the 1998 film, *Wag the Dog*, where a presidential administration simulates a war to distract public attention from other issues. Electronic information sources were not created to address a lack of information, then, but to create new categories of political and entertainment phenomena. Similarly, it is difficult to argue that the introduction of private transportation solved a transportation problem—the evidence suggests that automobiles have created a transportation problem that did not previously exist. Today, the average American must travel fifteen to twenty miles to work over a distance very often not serviced by any mode of transportation other than the automobile. Similarly, it is difficult to argue that the introduction of Internet access into middle-class homes addressed a need to spend more time in front of screen-displayed information and video games. Consumer societies arise after basic needs have been fulfilled and proceed thereafter to create additional necessities.

Ivan Illich finds the automobile exemplary of a contemporary condition in which "needs have become almost exclusively codeterminous with commodities. As long as most people walked wherever they wanted to go, they felt restrained mainly when their *freedom* was restricted. Now that they depend on transportation in order to move, they claim not a freedom but a *right* to passenger miles. And as ever more vehicles provide ever more people with such 'rights,' the freedom to walk is degraded and eclipsed by the provision of these rights. For most people, wants follow suit. They cannot even imagine liberation from universal passenger-hood, that is, the liberty of modern man in a modern world to move on his own" (13). Applying Illich's logic to information technologies, we can see the point where we can no longer imagine liberation from universal "receiverhood"—or from this new status as information receptors. The introduction of information technologies results primarily in a dependence on them and their codetermined commodities, such as hardware and software. Similar to the way in which automobile transportation shrinks the range of acceptable walking distances, information technologies shrink the range of acceptable duration or the recognized expanse of the now. Transportation and information technologies are thus properly understood as creating needs and wants by restructuring the environment in which human beings operate. Technology is neither bad nor good—it is not a moral issue but an environmental one. It is one thing to recognize how consumer goods are marketed, another to acknowledge their actual effects. Information technologies are sold as time-saving devices, but their introduction into middle-class households has contributed to the twentieth century's general attenuation of present time.

Colonial revolution in America made subjects into citizens, marking a political transformation; industrial revolution made workers into consumers, marking an economic transformation. At each stage, there are those who mourn the passing of old forms and those who welcome the creation of new structures of thought and behavior. Today, the revolution in information technologies will result in a parallel transformation of social definition, this time having to do with distraction, transforming audiences into information processors. People no longer simply receive information but are part of an information ecology whereby they consume great volumes of information created by others and in turn create information for consumption in the form of E-mail messages, chat-room participation, and web-page construction. People once sought distraction in the form of civic actions, literary encounters, or public events. Now, distraction has agency and is electronically enabled to seek out minds via privately owned technology and mass media. The information revolution plugs audiences into electronic sources that operate like information-stocked convenience stores, creating a need for continual updates and upgrades. In the same way that the automobile gave rise to the need for round-the-clock access to bread, gasoline, and food, the home PC creates the need for daily E-mail correspondence, fast and regular access to web-site updates, and on-demand access to on-line news consumption. This is the information environment. The issue is not whether this new social environment is evil, beneficial, or some corporate conspiracy, as we are not dealing with a moral process that can be halted by the better angels of unplugged existence. Rather, the question to ask is how the environment alters human consciousness in the form of needs, desires, and cognitive structures and how the mind might negotiate control over its processes in an ecological system that by design attempts to limit such control. We are clearly distracted by so much available information and so many messages. But can we exert control over our own distractions? Exactly how does "you've got mail" direct our attention?

Inevitably, some people will see the various information technologies as an aggregate evil, and these voices will join a respectable tradition in which new communication media are largely damned or accused of leading to intellectual degeneration. The list is extensive. For example, in ancient Greece, Plato attacked the written word for its degeneration of memory and attacked poetry for distracting minds from the hard questions of philosophical reality. The poet Samuel Taylor Coleridge (1772–1834) denounced the novel as an inferior and deleterious imaginative structure, contributing to a suspicion of that genre that lasted throughout the nineteenth century. Nineteenth-century physicians asserted with confidence that journalism, because of the way it presented public issues, making so many disturbing facts and contending opinions available, would bring on neurological disorders. In 1919, *Education* magazine condemned the cinema as a sure source of educational decline, a sentiment echoed continually in the 1920s, most notably by *Christian Century* in 1930. The *Saturday Review of Literature* led a crusade against comic books in 1948, accusing that form of contributing to a general decline in literacy. Finally, contemporary educators condemn the effects of television, rap music, video games, and Internet addiction on today's youth (Starker 7–9). Indeed, there has never been a new medium of human expression that was not, on its introduction, condemned by some responsible voice or some trustee of consciousness, depending on one's stake in the matter. Nonetheless, each successive condemnation has contained elements of truth, and human cognitive structures have indeed undergone stages of transformation whereby previous modes of thought are discarded. Plato was correct: few people today can memorize thousands of words, and in the great contest for loyal readership, comic books have proved more successful in this later phase than the *Saturday Evening Post.* At the same time, however, the mind continually accomplishes, withstands, and creates experiences that were impossible in the past. Right-clicking and Internet searches come easier to those born after 1985 with mouse in hand. Older generations object most strongly to changes in the world they witnessed in its—and their—formative stages.

My grandmother had a telephone table. It was a small wooden table, more like a desk, with a seat attached to it, and below the desktop was a small shelf on which she kept the phone book and her personal address book. This table was in tucked-away corner of the dining room; many households that preceded the telephone had such a furnishing. When she would make a phone call, she sat for a while at the table, thought about the call she was about to make, and then started the process of dialing (not punching) the numbers. And when the phone rang, every activity in the house ceased. If we were at the dinner table, all eating and talking came to a halt. "It's the phone!" someone would say, in the 1950s version of "you've got mail!" I can still see my grandmother straightening her dress and touching her hair before picking up the phone. I cannot imagine her sitting in the backyard in her bare feet on a portable phone, much less in the car talking cellular. Little did we know that in the struggle between the telephone and the dining room, the telephone would prove victorious and that the dining room, in many homes, would be given over to the household PC.

According to Steven Sarker, reaction to the novel, newspaper, comic book, cinema, and broadcast media has been "surprisingly consistent over the last few centuries" and marks a gradual shift in class-specific access to cultural sources. Sarker describes the historical pattern: "once any new media form or application achieved broad enough audience, it was perceived in certain quarters (e.g., intellectuals, clergy, educators) as a menace to the public health and welfare. Self-proclaimed experts quickly condemned it as preoccupied with violence and/or sensuality, overly graphic in form, and overpowering in its effects upon the young and the innocent. Its influence upon literacy, as well as physical, mental, moral, and spiritual health, was deemed unwholesome, and society itself was proclaimed at risk. . . . In each case, however, the widely predicted fall of literacy, sanity, and civilization somehow failed to materialize. In fact, as each media threat was replaced by a new one, the older forms became ever more accepted and respected" (143). Starker thus sees such negative reactions as a form of elitism, where the cultural intelligentsia fears a loss of control over mass audiences. "As each new media development blossomed and became yet another vehicle for popular culture, the elite became more and more removed from social power and leadership" (168–69). Whereas elite voices reach a very select and privileged audience, mass culture technologies are designed to reach and influence millions of people. New media forms are traditionally embraced by lower classes (as entertainment and an easy escape from drudgery), where trends emerge, while elites tend to cling to media over which they have achieved mastery. "As the mass audience broadened still further in the twentieth century . . . an increasing portion of the middle class experienced a threat from 'lower class' influences" (171). Disdain for such forms was the mark of membership in an intellectual and social elite. Nonetheless, information technologies and information-based entertainment industries have most quickly developed and spread throughout populations in affluent democratic cultures, especially in the United States, where respect for elite authority is perhaps the weakest of any democracy. Each development in this technology has made the reception and transmission of information more accessible to greater numbers, thus driving the level of public discourse closer to the level of the "people" than to elite segments of the population.

Internet proliferation is a particularly good example of popularization in the information age. Previous information technologies could be controlled by elites or at least by large corporations, because of the production expenses demanded by film, radio, and television. The cost of access to Internet technology is falling steadily, and the presence of a PC with Internet access will soon be as common in an American household as a television set, and the two media will in all likelihood be combined. However, because Internet access enables people to produce as well as consume information, the popular role of information processor reaches its historical culmination in front of the Internet-ready screen. Access to corporate, government, and mass media sources is within keystrokes. Web-page construction allows for the posting of any information or opinion without the intermediary of publishers. Elaborate sites are already devoted to conspiracy theories, various forms of shared interests and dementia, and, of course, pornography, which constitute the most popular Internet destinations. The expansion of mass-media reception and access throughout the twentieth century has meant that elite groups have gradually lost control over mass culture in America, over what constitutes a legitimate distraction from toil. It is unlikely that mass culture and manners are any less sophisticated than in previous historical eras. However, because mass culture may now be marshaled, marketed, and sold to millions of common people, it is both more profitable and more salient than elite culture. Selling PCs to historians to gain access to various research databases is far less of a market than targeting PCs to teenagers who want to visit fan-club web sites daily, purchase CDs, and send E-mail to their friends. Of course, those same teenagers will be able to participate in MSNBC's political-opinion poll and thus contribute to the shaping of public opinion and the lessening of elite control over public space.

Whereas populist critics like Sarker are amused by elite condemnations of mass culture, elite critics such as James B. Twitchell are concerned that the level of cultural discourse is in rapid decline in the information environment. Noting that the category "vulgar" has disappeared from cultural commentary, Twitchell argues that contemporary Americans thus "no longer believe in the category of taste" but are inclined rather to canonize the vulgar in popular culture (18). The normalization of the vulgar is symptomatic of a drop in "the center of gravity—the 'norm'— in Western culture and world culture" (23). The drop is a direct result of an increased access to the reception and production of information and entertainment. "As the economies of mass production give greater access to those previously excluded—the young, the unsophisticated, the aggressive—the stories demanded and produced become progressively more crude and 'vulgar'" (260). Twitchell is well aware that vulgarity has a long history in the United States and Britain, including Elizabethan freak shows; popular carnivals in the eighteenth and nineteenth centuries; Punch and Judy shows, cockfighting, dogfighting, bull-baiting, and games of chance in the nineteenth century; and, of course, P. T. Barnum in the early twentieth century. Pit-bull dogfights and cockfighting continue in contemporary America as well. The vital link between popular culture and vulgarity is not without its genealogy. What we witnessed in the twentieth century was the democratization of culture, so that the term—*culture*—is no longer associated with elite or instructional sources but with the common people, their concerns, their preferences, and their distractions.

If you set your web browser to www.jennicam.org, you will arrive at JenniCam, an Internet site devoted to recording the life of a woman called simply Jennifer, a college graduate from Pennsylvania who lives in Washington, D.C. Her web site is devoted to chronicling her life, including her daily (and nightly) activities. She produces the *JenniShow,* which is an occasional, low-budget webcast production about such things as visiting the airport and showing off her wardrobe. An on-line, digital camera is permanently set up in her apartment, and for a modest subscription fee, Internet users may have a twenty-four-hour window into her private space. Visitors to the site may gain free access to the live camera in twenty-minute, still-photo updates. JenniCam (and numerous sites like it) marks the triumph of the vulgar, in the truest sense of the word, referring to the behavior, language, and interests of the common people. JenniCam makes a cultural production out of the life of lowly, vulgar Jennifer and does so without production companies, marketing strategists, or commercial network services. Instead, a mundane woman with HTML training has become famous for being completely ordinary, except for the fact that her life is being digitized and made available as information for anyone with Internet access to process. JenniCam marks the triumph of mass culture as a mirror of the common lot of Americans, who, when they seek entertainment, seek what all social classes crave: a mirror to normalize their own beliefs, fears, and desires. Jennifer's viewers, like Jennifer, are information processors, receiving information about her life and including that information among their daily distractions. Jennifer claims to respond to thousands of E-mail messages daily, which JenniCam incorporates organically, as it were, because reading the mail is a part of Jennifer's digitally chronicled life. Thus JenniCam viewers are among her daily distractions, which, for Jennifer, constitute life as she has constructed it.

At the same time, as information technologies play a greater role in the intellectual environment and as the digital age becomes organic to human thought processes, cognitive patterns will inevitably be called on to adjust. Younger minds, whose patterns of cognition are still under construction, will evolve as a matter of development, as will minds of all ages not heavily invested in outmoded or threatened structures of thought and behavior. As always, some minds will find the new cognitive demands threatening and will resist change in the name of a simpler, more familiar, and thus more valuable past. This progressive pattern has been constant, as minds in firm possession of established cognitive structures have consistently objected to new technological demands. Younger minds have with equal consistency attached their development to new cognitive structures, if for no other reason than to lay claim to an emergent world being abandoned by their elders. Resistance is particularly vehement by those who have mastered an established form, as in the case of William Faulkner's famous refusal to allow a radio in his house. With each new technological form, the mind has been compelled to expand and alter its patterns of cognition. Historically, the results are unpredictable and erratic because human beings continually introduce alterations to their environment that, in turn, demand adjustments to cognitive structures. The attention of the mind today is on information; and like transportation at the beginning of the last century, we cannot seem to get information quickly and conveniently enough to suit the needs of the environment of distraction we have created.

One distinguishing feature of contemporary culture in America is that information is consumed for diversion rather than for consequential purposes. Paradoxically, in other words, information of all kinds is processed as low-level cognitive distraction, a form of mass entertainment. The collapsing of news, data, broadcast, and leisure media has resulted in an entertainment culture where the act of consuming information is as diversionary as watching a variety show or a situation comedy. The reasons for this development are not surprising. It is satisfying to be under the impression that one is successfully processing information that, if not for the watching, one would be missing. The information media share a universal marketing strategy, inherited from newspaper sales, of headline and special-edition presentation, always designed to maintain the illusion of urgency. The question, of course, concerns the effect of the information being consumed on the watcher. It is quite easy to be satisfyingly distracted by the illusion of learning something important. However, as it turns out, it may feel as good to gain insight into the nature of the human predicament as it does to learn that some man from Oklahoma has discovered that his wife has had affairs with various delivery men, all of whom are waiting offstage to tell him about it. The satisfaction of knowing, that is, may feel the same whether one knows the origins of the cold war or the story of the man who married his horse. The simple term *information* may no longer be adequate to describe the spectrum of input available to human beings in their capacities as data processors. We live today in an information wilderness, where it is not always so easy to find a path to knowing something of consequence.

A growing percentage of television programming and a good deal of Internet content consists of information packaged and designed to entertain by informing, whether in the manner of newscasting, consumer advertisement, celebrity gossip, or creative programming with a specific social or cultural agenda, such as movies about rape victims, cancer patients, or other traumatizing conditions. Perhaps the best example of information packaged as entertainment, also called *infotainment,* is the immense popularity of the Weather Channel, also available at www.weather.com. On the cable-TV Weather Channel, viewers are entertained by extensively detailed weather reports on all regions of the nation and brief reports on international climate changes. Special reports are created for dramatic weather like hurricanes and blizzard-level snowstorms. There is no practical reason for a sedentary viewer in New England to know the forecast for a hailstorm in Mississippi; the information is completely inconsequential. Nonetheless, it is an instance of accurate information available for processing, and the consumer of it feels as if she has been informed, senses that she knows something that, had she not been watching, she would not know. It feels good to be in the know, a fact that explains the popularity of most infotainment, from talk shows to newsmagazines. The marketing of these programs creates an illusion of the imperative ("don't miss this"; "an exclusive story"; "a behind-the-scenes look"), with teasers issued regularly to create and maintain the ongoing sense of urgency generally associated with the medium of television.

Entertainment has been big business in the United States, and information is now packaged and sold in the same way that the circus was marketed, the same way that soap and automobiles (and for some time now, politicians) are marketed. And whereas in one era the circus would come to town and distract citizens from work, school, and various other obligations to make people make time to attend the ring and tent, today information (as entertainment) comes into our household, our automobile, our workplace, and onto our word-processing screen as a convenience, to distract from whatever we should be doing but prefer not to be doing, to tell or show us something that may capture our attention for a little while and (not incidentally) remind us of consumption opportunities. The processing of information thus functions as a manifestation of the contemporary American consumer-ideological structure. It is not the specific content of any particular set of ideas that informs this peculiar ideology but the simple presence of information for consumption. Americans are thus under the impression that there is an explosion of information available for processing and that almost no one has enough time to consume it all or even to be aware of all of it. Those who market new technologies support the supposition by continually claiming that information technology exists for the convenience of busy people. News reports are broadcast repeatedly and repetitiously for the benefit of those who are reminded that they are too busy to attend to the processing of such information for sustained periods of time. We know that the accumulation of conveniences throughout the twentieth century made us feel as if we had less and not more time—but such knowledge is impossible to assimilate without critically interrogating the effect on cognition brought on by the collapsing of information and entertainment.

When you learn to drive you have to learn what to attend to and what to dismiss, what to focus on and what to allow to blur into the background unless needed. A decision to watch the rearview mirror for an extended period of time would be disastrous, for example, and resting the foot on the brake would wear away the linings. The first few miles of driving bring an onslaught of sensual and intellectual stimuli. The car seems impossible for the neophyte to steer in a straight line; corners appear to be nonnegotiable; sitting on the left side of the vehicle makes the edge of right side indecipherable. And then there is backing up, parallel parking, and U-turns, a series of navigational impossibilities. If we think back to the constant challenge of processing the information necessary to drive in a straight line at seventy miles an hour or to negotiate through traffic with moving obstacles on all sides of our vehicle, we find a model, in microcosm, of the management of distraction called on by the information environment. After a few months, it's easy. The phenomenon is how few crashes there are. Driving means negotiating through endlessly generated distraction, learning how to glance, and expanding one's sense of the space one occupies while in motion. The information environment demands a similar set of skills as we sit, immobilized but poised to receive stimuli. We are not told where to focus but simultaneously given hundreds of options. The need to negotiate distraction is imperative when everything before our minds beckons our attention and there is nothing, nonetheless, that we really need to know.

It would be a serious intellectual mistake to confuse information that functions as entertainment with actual, or knowledge-based, information. It would be a mistake as well to simply ignore the cognitive implications of information processing as entertainment. Real information, such as who controls wealth and property in the United States, why prison building outdistances school construction, or comparative rates of upward and downward economic mobility, is as difficult to locate as it ever was and must be culled from the kinds of books and journals not featured and sometimes not even carried by the megabookstore at the strip mall or reported on by television features. The same is true of knowledge, the kind of understanding gained by sustained research and meditation into a subject area or the deeper insights into the human condition gotten from complex literature, neither of which is available as simple information, although most of which can be reduced to informational statements. It is true in a way, after all, that the meaning of *Oedipus Rex* is "shit happens." The naughty phrase is entertaining, as is most information packaged in that manner. However, as is often the case with aphorisms, the succinct phrase may be used as a substitution for comprehension. Real knowledge continues to exist in contemporary society, and it too consists of information but is not dependent on its commodified form for its truth claims. There is a profound truth in Genesis concerning forbidden activity, the emotion of guilt, and the sense of God's existence, for example, that survives the debunking of the creation myth by scientific research. However, we are not talking about this kind of information, just as opera is not what is offered by daytime TV soaps.

Infotainment (and I would include TV newscasting in this term) offers the illusion of knowing in the same way that a television drama offers the illusion of reality. When we watch fiction on television, as drama or as situation comedy, we suspend our knowledge that we are watching simulacrum and enjoy the illusion of seeing into another set of lives and situations. Very few viewers confuse the roles played by actors on television with the lives of actual persons, although there are exceptions. However, when we encounter someone who considers a soap opera character to be an actual, consequential person, we are likely to be amused and to think the person naive or maybe stupid. Marketers of horror films often use the tag "it's only a movie" to assure audiences that the film will in fact be terrifying and may cause viewers to lose the distinction between real and entertaining terror. Terror that entertains is enjoyable because it is inconsequential. Similarly, when we watch the news, the TV newsmagazine, or the special report, we enjoy the illusion of gaining insight into some aspect of contemporary reality. However in the case of infotainment, the viewer is expected to maintain and not suspend a sense of the real. At times such maintenance may be difficult, especially when the report is about something with which the viewer has some expertise. In these cases, instances of bias, fabrication, selectivity, and error are recognized. If the reported incident happened where you live or work, for example, you will immediately see how the story has been constructed toward some specific agenda. Most infotainment, however, centers on materials otherwise inaccessible. Local news, for example, concentrates on crime—an area of information on which the average middle-class viewer, protected by social standing from such affairs, cannot perform reality checks. Familiarity rarely distracts. The things we know best we take for granted; desire, not satiation, fuels distraction. Celebrity gossip, which seems so vital, is equally remote. Whereas fiction as entertainment requires the suspension of one's sense of the real, infotainment requires something entirely opposite, the mainte-nance of one's sense of consequences despite the inconsequential nature of nearly all of its content. There is a vital difference in the two experiences.

We are trained to process the dramatic illusion of reality by two thousand years of literary tradition. Schools teach English by the use of fiction and poetry and explore the relationship between language and reality to stress the distinction between the world of physical experience and the world of linguistic creation. When we watch the television drama *ER* or even the situation comedy *Seinfeld,* with its claim to be "about nothing," we readily test the represented reality against our experiences and judge the success of the representation according to the nature of the relationship between the illusion and the experience. If *ER* represents the world of the medical emergency room satisfactorily, we enjoy the data we receive from it. If *Seinfeld* successfully represents an insight into the young urban middle class, that projection of reality is enjoyed as well. However, the illusion of informa-tion is another matter entirely, one for which most consumers are ill prepared. Television drama tends to be about matters that can be tested easily: family dynam-ics, police and hospital matters, dating, seduction, fear, loneliness. Information broadcasting, such as the newscast, tends to be about unverifiable matters for the simple reason that the represented subject is purported to be actual rather than staged. That is, the crime scene depicted is the actual crime scene—where the crime was committed—not a representation or dramatization of the crime scene. If the infotainment material were readily verifiable, there would be no reason to watch it on television or find it on the Internet. One watches (or surfs) to find mat-ter that would otherwise be missed. Almost by definition, then, what makes good infotainment is unverifiable and must be accepted by a kind of faith in the integrity of the producers or sources. The reality of the murderer's murder cannot be tested without interfering with the police; the reality of the military action can-not be tested without placing one's life in danger. When the Weather Channel covers my region on its national map, however, it very often misrepresents the local situation in my neighborhood. When infotainment concerns us or matters we know a great deal about, we glimpse the tenuous nature of information packaged in this way.

The growing popularity of professional wrestling on television is testimony to the domination of the medium by infotainment structures. Broadcasts of live competitive sporting events (a category that does not include professional wrestling) have been perhaps the culture's best example of infotainment. Teams or individuals are matched in a contest where the results are unknown, and after an entertaining interim, information is gained defining the winner and loser. That information is then announced for hours afterward on news and sportscasting programs and then entered into a statistical data bank where the illusion of consequences is maintained in the way of world records, pennant races, and similar benchmarking. Participants are paid enough money to fund the illusion, and any argument that sports don't matter may be countered by reference to nonillusory, astronomical salaries and the importance of role models for young people. Viewers are encouraged to try the sport (usually as children), making the difficulties of skillful play widely known. For all of these reasons, competitive sporting games make for successful infotainment television, radio, journalism, and Internet sites. The idea of professional wrestling, then, should not work at all as infotainment. The contests are fixed, or staged, and are not competitive; the skills presented are not the skills watched (these men—and women—are not hitting each other but practicing falls and feints); and the drama displayed is entirely simulated. Why, then, is it so popular? The reason is quite simple. Professional wrestling is the perhaps only completely truthful infotainment broadcast today—truthful, that is, to the medium. It is the single infotainment event where viewers get exactly what they desire: wholly inconsequential but highly technical and dramatically packaged information. Competitive sports, while also inconsequential, carry the weight of urban, regional, and scholastic loyalties and thus may result in a genuine sense of loss or achievement. Professional wrestling drains every vestige of consequence from its competition and so may be enjoyed as pure—that is to say, always already obsolete—information.

Despite the difficulty of verification, information broadcasting is representational, not experiential. Accuracy is established more through formal, aesthetic gestures of objectivity (such as presenting "both sides") than by the viewer's ability to test the facts through experience. The impossibility of experiential verification results in a situation where the burden of credibility is not on the viewer's ability to check facts but on the veracity of the representation itself. Questions of authenticity are likely to be lost in the general stream of information transmitted or presented. Instead, infotainment simply has to ring true to be believed—true to ideological presuppositions, that is. In fact, the information received from broadcast or Internet sources far more closely resembles drama and sitcom in effect on viewers than is normally acknowledged. In both cases, an illusion of reality is evoked that substitutes for physical experience. Americans who have no idea what a middle-class lesbian lifestyle might entail learn a great deal from *Ellen*, a situation comedy that focused (at least in its final season) on a gay woman's experiences. Similarly, Americans who have no idea where or how international terrorists operate find out this information from news reports that offer expanded coverage of bombings or reprisals. However, again, the key difference is that one may test the accuracy or insights of *Ellen* against the lives of people one knows through experience and thus judge the representational biases of the television program. In such cases, the viewer quickly learns that the representation is particular and singular, not symbolic or universal. Not all lesbians are Ellens. Such critical judgment is impossible with terrorist coverage—or similarly, with reports of corporate mergers, government policy initiatives, scandals, and so forth. Infotainment offers an experience of viewing that feels like the process of knowing. As one would expect, *Ellen* was far more popular among people with no lesbian experience than with those familiar with such human variety. Informational entertainment involves the consumption of information and provides a substitute for thought while distracting the viewer from knowing anything that is verifiable experientially. As a result, one's control over the consequential level of intellectual distraction is abdicated.

Professional wrestling abandons all traces of verifiability. Not only are the contests fixed in the sense that the results are determined by producers, but the fighting itself is choreographed, a succession of elaborate lifts and falls that require hours of practice to create a convincing display and to avoid injury. Participants are provided identities by producers, and there have been incidents of actual confrontation when a wrestler's identity is unilaterally changed from "good guy" to "bad guy" or when victorious characters are transformed into losing contestants. The popularity of the wrestlers is determined by the reaction of the crowd to their personae and performances in and out of the ring, and the veracity of their representational mask is crucial to their career success. Furthermore, the dramatic nature of the enterprise is an open secret. Ringside announcers, magazines, and web sites devoted to the industry report on the sponsoring organizations—such as the World Wrestling Federation and World Championship Wrestling—as if they were competitive sports, and sold-out crowds attend events to cheer their favorites, as if watching a situation whose outcome was known by no one and as if a favorite wrestler's efforts were being expended to determine, rather than fulfill, a result. All involved in professional wrestling, from wrestlers to promoters, announcers, magazine writers, and fans, participate in an elaborate illusion of competition, individualism, and (sometimes) fair play. One might deconstruct the phenomenon and detect a critical interpretation of these values, where competition, individualism, and fair play are mocked by all involved. More likely, professional wrestling marks the maturation of the infotainment processor, the viewer who knows that little of what he consumes from such sources represents the truth of experience. Rather, the experience of viewing contains truthfulness—the forms and structures of the medium itself, not its content. The viewer desires the experience of processing information, freed from the consequential implications of knowing.

Experience is a vital concept here. A growing number of television, cable, and Internet industries provide the experience of information. When an area in which the viewer has competence is represented as information by any of these media, the representation will be tested against actual experience and judgment formed accordingly. However, when an area in which the viewer has no competence is represented as information, then the represented information becomes the extent of one's experience. When one enters the woods, the experience of trees, underbrush, bird and animal sounds, and sun or moonlight amounts to the environment one must traverse to successfully navigate a path and reach a destination. If the woods are familiar, departures from familiarity and normalcy become noteworthy—a newly felled tree, a dried-up creek, an abandoned bird nest. If the woods are unfamiliar, everything is noteworthy because everything is processed sensually and thus seems extraordinary. When one traverses the information environment, a parallel menu of environmental factors come into play. Because broadcast information depends on sensational, singular events for its content, most viewers remain lost in the woods of information and consider whatever is being broadcast to be newsworthy. If such were not the case, no one would seek out or attend to intrusive infotainment sources. For example, if information broadcasts consisted of examinations of various government offices and agencies not during crisis periods but simply to provide information about them, it would be difficult to convince viewers of the necessity of becoming distracted from other responsibilities to watch. However, should an office or agency become embroiled in a crisis, a sense of urgency is created that compels information processors to attend to the "breaking story."

One result of crisis-based information is that the world represented by infotainment appears eternally on the brink of disaster, with little time available for contemplation or the cultivation of real, consequential knowledge. Crises also contribute to the shrinking of the now, focusing attention on future developments and the results of confrontations rather than on processes and deliberations. Political incidents are hyped like wrestling matches, and the structure of infotainment programming requires headlines and up-to-the-minute developments. Thus the infotainment world seems inexhaustibly exotic, bizarre, and threatening, a situation that has alarmed cultural analysts. "[T]his equation of experience with the exotic is a disaster, psychologically, morally, and philosophically," according to Edward S. Reed. The equation impoverishes daily experiences and trivializes events at both local and global levels. "If experience can be gained only in unusual situations, it becomes the preserve of the hero or the specialist, and wisdom is transformed into mere expertise" (21). It is as if we were lost in the woods with a guide who insisted that the woods were eternally unknowable, in crisis, and subject to incessant change. The guide, the information environment, offers the continual experience of information but the sensory media on which it depends—visual and aural, primarily—is incapable of providing an arena for the experience of deeper, thorough knowledge acquired through sustained cerebral activity. The infotainment vision of the world implies that deeper thought—an appreciation for complexity and a rejection of the facile (but entertaining) detection of crises— is a waste of valuable time, time more appropriately spent invoking and covering the steadily unfolding (and always changing, we are told) crisis.

Again, the reason is simple. Reaching a point where contemplation is necessary, where local, unplugged knowledge may be cultivated, would mean disengaging from the infotainment source. Once a person ceases to process information in this environment, she becomes unproductive, a deadbeat consumer. It is, to return to an earlier analogy, like learning how to drive. Once we accustom our minds to the space occupied by the vehicle, to the responsiveness of the pedals and of the steering wheel, then we no longer need the guide or the instructor telling us where to place our hands and how to center the automobile. But imagine that the automobile came with the driving instructor permanently installed, saying "turn left here" and "go three blocks and parallel park." Our driving experience would thus be eternally tension filled, and we would never know exactly were we were going. On the contrary, we would never *go* anywhere except where the instructor directed, and the only reason we would drive would be to be observed and charted. The information environment functions like one massive driving lesson, with voices and other signals continually directing our attention, testing our responsiveness, while purporting to give us what we need before we ask. At some point, however, the mind must disengage the voice of the instructor and assume authority over the technologies of information. To drive one must master the distractions that inform the experience. To survive the Age of Distraction, one must likewise wrest control over one's attention. The Information Age needs to be unmasked, in other words, and the instructor expelled from the vehicle.

Various media refer to the present era as the Information Age. The designation is clearly meant to imply and describe a progression from such earlier eras as the Industrial Age, the Age of Discovery, the Age of Reason, and similar historical demarcations. What is pointedly evoked is a change in human consciousness equivalent to the restructuring of cognitive modes that followed the assaults on faith by rationality, the expansion of the known physical world by sea travel, and the transformation undergone by human beings from producers to consumers as a result of industrialization. The Information Age thus signals an expansion of the metaphysical world, including the extension of the processes of commodification beyond goods to encompass ideas, data, facts, fictions, and all forms of cognitive input, grouped generally within the category of information. Human beings have created a succession of environments to house and accommodate their needs— agricultural, industrial, communal, private—and each successive, parallel environment has introduced changes into the cognitive modes employed to comprehend existence. In the industrial era, cyclical time, associated with an agricultural ecology, was augmented and largely supplanted by linear time, associated with historical thought and industrial production. Similarly, the Information Age demands new cognitive modes capable of benefiting by and coping with the massive quantity of intellectual stimulation currently available to human beings. No longer something sought after when needed, in the way a person seeks out a plumber when the pipes leak or consults the manual when the machine malfunctions, information in the Information Age is environmental, it signals life and vitality in the home, in the workplace, and in public spaces such as street corners, waiting rooms, and airport terminals.

Robert Toll, in his history of American show business, points to the way technology has transformed entertainment, including "the transformation of the average American home into an amusement palace, a fundamental change that had deep implications for American family life and for American business, which sponsored broadcast entertainment" (4). More than an amusement place, today the home is also an infotainment center, where Internet technologies place unlimited database sources at the hands of entire families and Direct TV makes available virtually every television broadcast in the world. Home computers can be programmed to monitor changes in web pages and to visit sites on a regular basis; VCR technologies may be programmed to watch and record television content at all hours. The marketing strategy in these matters is to give the information consumer "control" over data and entertainment sources and, as a result, increase one's sense of the necessity of continually processing information. In addition, information does not have to be sought after because the relationship between it and human consciousness has been reversed. Now, information seeks out the conscious mind: when information distracts, when it captures the attention, its purpose is accomplished.

Of course, information secures attention not by claiming to distract but rather by claiming to redirect, divert, or entertain. The working assumption behind information technology is that it provides a service to the clueless, the bored, or to those defined as needing (or possessing the right) to know something, anything. In the airport terminal, CNN entertains passengers with news programming designed for travelers. Prior to television monitors, individual passengers read books or magazines or watched airplanes or each other. Now, groups of passengers watch the television, featuring the airport network produced by CNN or a local broadcast station. The experience of broadcast information augments the inescapable environment of anticipation, departure, and arrival in the terminal. Whatever thought was induced by the private mind interacting with its preferred stimulation is replaced by the service of infotainment. At home, twenty-four-hour news networks repeat the most superficial coverage of sensational news stories (and leave unreported the workings of government and finance capital) while constantly teasing viewers with the next breaking story or the appearance of someone with expertise or inside information. Consumer marketing interrupts televised programming, while television programming itself includes increasing amounts of informational product, from talk shows to celebrity gossip, televised newsmagazines, special reports, and the like. Raw information programs, such as those where police business is televised, eliminate the news mediator and bring infotainment directly to the human processor. In all these cases, the illusion of knowledge is gained by the consumer, who in all likelihood is unaware of being distracted from knowing or thinking about much of anything except the need to keep watching. Internet browsing technologies take the process a step further, so that nearly every site on the World Wide Web contains links to other sites creating a universal experience of redirection, diversion, and infotainment.

In 1958, Aldous Huxley noted "man's almost infinite appetite for distractions." Early in the Information Age, Huxley voiced his concern that an environment where most people in society "spend a great part of their time, not on the spot, not here and now and in the calculable future, but somewhere else, in the irrelevant other worlds of sport and soap opera, of mythology and metaphysical fantasy, will find it hard to resist the encroachments of those who would manipulate and control it" (46). Huxley's warning is made more relevant by the addition of information sources that claim to add reality to the worlds of sport and soap opera by augmenting fictional or performance diversion with sensational infotainment or the illusion of responsible, engaged distraction. Nearly all information and infotainment media sources mask most successfully their function as corporate interests and their actual relationship with viewers—or more accurately, the consumers of their product. The business aspect of mass media is directly responsible for the nature of television information. The primary mission of broadcast, cable, and Internet information providers is to distract consumers and instill dependency on the sources of information provided. Infotainment sources do so by commanding a sense of urgency that results in the consumers' inability to disengage. If one is on the verge of finding out something of value, one is less likely to turn off or disconnect from the source.

In one sense, infotainment sources operate as gossip technologies. Gossip is pure information, sourceless ("have you heard about . . .") and humanly irresistible. On the Internet, there are many sites like JenniCam, where cameras are set into the homes of persons who simply go about their lives or go about lives designed to provide hours of distraction for Internet viewers. To produce these webcast programs there is no camera operator, no director or human eye selecting camera angle or subject. Instead, the PC simply collects data, embodying what Paul Virilio has called the "vision machine," the machine that supplants or substitutes for the human capacity to view or choose what to look at. New technologies are thus "preparing the way for the *automation of perception,* for the innovation of artificial vision, delegating the analysis of objective reality to a machine" and removing the human element from the production or maintenance of fields of perception. "This is the formation of optical imagery with no apparent base, no permanency beyond that of mental or instrumental visual memory" (59). The implications for our understanding of information are profound, as the link between information and source is dissolved when information is gathered mechanically rather than humanly. The human desire for gossip is thus commodified and mass-produced, as gossip possesses no point of origin, no responsible agent of instigation.

The vision machine commodifies the familiar gossip structure so that the ritual opening, "have you heard about," is safely removed from any danger of accountability. Who told you? The camera told me. Infotainment operates in a cosmos whose entire reason for being might be summed up by the prefaces "have you heard" and "have you seen." The continuum of webcast lives and talk-show confessions may be extended to include investigative reporting, live "cop" shows, and those ubiquitous "sources" of high-level information who are quoted but never named on the evening news. Literary scholars have talked about the death of the author for a generation; it may be time to add the death of the source and origin. The "vision machine" is simply the reification of a structure already established in the mass media, where anonymous high-level sources compete with anonymous guests on confessional talk shows for the attention of minds looking for the latest datum. Information, as infotainment, exists independently of origin and receiver. It simply is. No longer understood as the product of human instigation and thus implying human responsibility for its direction and use, information has become a self-perpetuating commodity. Information exists independently of source and destination, and thus it may be traded, bought and sold, and then junked like an abandoned car once it is found to have no use in distracting someone, somewhere, from self-directed thought.

The evolution of the terminology itself parallels the phenomenon under examination. The changing definition of *information* reflects the progression of the technology that has developed to mass-produce it. Early usage of the word, dating to the late medieval period, referred to the effect of information on its recipients. Hence, *information* was understood as something that acted in the "forming or molding of the mind or character"; it was understood as an element of "instruction," for the "communication of the knowledge of some fact or occurrence." If an item learned was of no consequence, it was unlikely to be considered information in this early definition. However, from sixteenth through the nineteenth century, the definition of the term became less rooted in the source or the recipient and was understood as "an item of news" or "an account." The link to "forming or molding" the mind was shed, and the definition began to hinge on the qualities of the phenomenon itself. By the nineteenth century, the definition of *information* simply depended on whether it was newsworthy or explanatory—presumably to someone. However, in the twentieth century, according to the Oxford English Dictionary, *information* may be understood as a phenomenon wholly divorced from producer, audience, or consumer. The OED defines twentieth-century usage of *information* as "Without necessary relation to a recipient: that which inheres in or is represented by a particular arrangement, sequence, or set, that may be stored in, transferred by, and responded to by inanimate things." Agency and reception are not necessary to this definition; information has become an environmental phenomenon.

Thus, with vision machines in mind, the latest stage in the evolution of the human conception of information is that information exists independently of origins, without necessary relation to a source. Questions regarding the extent of human agency, central to philosophical ruminations about postmodernism, for example, are surely tied to the absence of agency associated with the production and dissemination of information. As an alternative model for the question about the felled tree in the forest, information exists whether one sees, hears, or consumes it. With automated cameras (some hidden and some not) mounted on police cars, in workplaces, throughout the home, and wired to the household PC to webcast images on personal home pages, information is quite accurately understood as both sourceless and destinationless, forming an actual—not a virtual—environment. In fact, the term *virtuality* may pass from use, like the term *horseless carriage,* signaling a period of technological transition and the residues of an earlier conceptualization. Virtuality marks the present evolution of information from recipient- and source-bound phenomenon to independent, agentless phenomenon to environment. Human beings have learned to live and be productive within steel-girder and concrete structures and in controlled climates. What faces us now is survival within electronic information environments.

Cultural critic Cecelia Tichi has provided a thorough account of the alterations in cognitive modes stimulated by the incorporation of television technologies into local, private human environments in the United States in the 1950s and 1960s. Specifically, in this period, television became a functioning part of many, if not most U.S. households. While the television may not be watched all the time, it is turned on for most families' time at home. Few minds raised on television are incapable, though some may be unwilling, to divide their attention between the physical household and its televised reception content. Television enables and demands "a new pluralistic state of consciousness in which engagement in the on-screen realm proceeds simultaneously with engagement in diverse activities in the habitat. Attention shifts to and from the television, including it but not excluding other activities" (119). With the introduction of Web TV and PC-mediated environments, including the eventual merger of Internet, cable TV, and telephone technologies, Tichi's notion of "teleconsciousness" becomes increasingly vital. "In this formulation, thought itself changes, becoming multivalent. The individual is cognitively functioning in two or more places simultaneously, in affairs of the habitat and in those of the on-screen world, assigning primary attention alternately to one or the other" (120). Thus, consciousness confronted by television and interlocked with the information environment generally becomes multicentered, chronically distracted to the point that distraction is its chief characteristic. Or, as Tichi has it, consciousness is marked by "a continuous reassignment of priorities of attention" (121). What unifies this intellectual structure of distraction is the continuous search for the experience of information, the defining experience of the media environment. The experience of information is not dependent on content but on the need of the teleconscious function of the mind for plurality.

Far too much media criticism has centered on content. Twitchell asks instead that we turn our attention to studying "the experience of television" (203). Viewers crave the experience of television, just as the experience of Internet surfing attracts a new generation of viewers to an advanced information technology. As for television, the demarcation of eras according to television series, as opposed to single programs, bears out Twitchell's thesis. A popular, long-running television series (*All in the Family, Dallas, Seinfeld*) encapsulates the experience of watching and gathering information over a specific period of time. In syndication, these programs remind the viewer of another time period, just as returning to one's high school reminds one not so much of algebra and English but of adolescence. One remembers content far less than the experience of certain processes when one makes such returns. A contrast may be drawn between such experiences of nostalgia and the act of reading. While one may associate certain reading experiences with particular points in life—adolescence and *The Catcher in the Rye*, for example— rereading favorite books does not mark a nostalgic return to the point of initial reading. On the contrary, rereading *Middlemarch* plunges one entirely into another conscious world, and while the reader may recall the first time she read the book and compare impressions, the text's power is such that it will fill the conscious mind sufficiently to become its own experience of reading. It would be like going back to high school and on hearing the bell, attending algebra class once again.

Debates over the relative and comparative experience of reading and watching television—and now, reading and Internet surfing (with its illusion of reading)— are solidly embedded in the public culture. David Marc summarizes the impasse well and describes its effect on children, on whose lives the debate is usually centered. Educators agree that Johnny can't (or won't) read. The lines of contention are between those who fight TV by arguing for limiting viewing time and those who say we ought to bring TV methods into the classroom, to "technologize" literacy education. More likely is that neither method makes much sense and neither approach will succeed. Parents who sit in front of television sets and tell their kids to read more are not likely to have the intended impact. Teachers who bring television sets into their classrooms to make reading more fun may experience the same disappointing result. According to Marc, "[C]hildren are quite possibly being driven to further distraction by the mixed signals that result from this squabble among adults." As children of infotainment culture, they have to watch TV to process sufficient amounts of information and to avoid the depression that follows the stigma of being out of touch. "However, they also know that too much TV will lead them to fail in the tasks they must master to get what TV says they'll get if they are successful" (133). The issue is not TV versus reading (or TV versus playing outside or TV versus doing homework) but understanding the function of television in the information environment. I suspect that the same if not a greater percentage of people are reading seriously today as have ever done so in the modern era. The difference is that reading, which has remained a constant experience for centuries, exists in an unprecedented environment of information-processing opportunities. To judge that environment as competing with reading is to vastly underestimate its importance and ubiquity, similar to seeing the Industrial Revolution in competition with gardening. Educators need to explain how the information environment is shaping modes of consciousness so that their students will be able to more satisfyingly navigate within it.[1]

We should know that what Tichi calls the "continuous reassignment of priorities of attention" (121) signals the presence of chronic distraction. In the past, one had to exercise some initiative to find intellectual distractions—a book, a sermon, a film, a public or civic program. Today, intellectual distraction is channeled into the home. Critic Raymond Williams first established the description of the television experience as being fundamentally the "fact of flow"(95). Television watching is nonsequential, and even if one program is watched, as opposed to channel surfing, the viewing is interrupted continuously by commercials. More accurately, though, people watch television while their teleconscious minds are also, perhaps, reading magazines; having sporadic conversations; preparing, eating, or cleaning up after a meal; or viewing a number of programs simultaneously. Williams describes the process of one day's television watching as "having read two plays, three newspapers, three or four magazines, on the same day that one has been to a variety show and a lecture and a football match. And yet in another way it is not like that at all, for though the items may be various the television experience has in some important ways unified them" (95). The unifying experience lies in the act of information processing, the sense that television provides the viewer with what should not be missed. Consuming information is the cognitive demand made by the environment of television, regardless and unrelated to the specific content, quality, or intellectual level of the information in question. Subsequent technologies only reaffirm Williams's thesis, as these technologies are marked by a flow into which the web surfer interacts, clicking on links, creating his own rhythm where TV had previously managed the rate of his infotainment experience.

The environment of information encompasses all media technologies, from television newscasts and situation comedies to www.msnbc.com and www.jenni cam.org. The most popular situation comedies have brought civic and public information to American households, informing mass audiences of how particular classes within society live and work, how single women raise families, how liberal children interact with less liberal parents, how black families differ (or do not differ) from white families, how races and sexes interact. Similarly, the most popular dramas have brought information about various professions and economic classes to mass audiences. It is in the nature of the medium for television to collapse information and entertainment into one continuous flow of stimulation and distraction. Internet technologies expand the sense of consumer autonomy by allowing viewers to customize their screens and receive information they choose, from local weather and sports news to personal stock reports. While some of this information may have consequences for the consumer—if it may rain, one takes an umbrella; if personal stocks fall, one may wish to sell—the vast majority of it is highly satisfying to consume because of its inconsequential nature. In general, it feels good to learn something, especially when it can be learned without difficulty and if it does not inhibit directions and priorities already in place. Information makes highly effective entertainment when it doesn't make any difference to the life of the person processing it, the person who has the pleasure of knowing the scoop, the news, or the gossip. Tragically, though, the illusion of being informed often substitutes for the possession of consequential knowledge, and the arrogance that accompanies the illusion may impede the processing of complexity.

In first three quarters of the twentieth century, the business of entertainment was limited to the manipulation and projection of print structures, such as drama, fiction, screenplays, and wire-service information. The end of the twentieth century marked the beginning of what many have called the Information Age because of the apparent explosion in information available to human beings. However, the age is marked not so much by consequential information as it is by the commandeering of consciousness—information as entertainment—and the consumption of information as a way to satisfy the insatiable human need for distraction. Useful and consequential information is as difficult to obtain today as it has ever been. In fact, information with consequences may be more difficult to encounter because of the illusion of information's accessibility. Entertaining and distracting information, conversely, is readily available by cable television, Internet access, or voice-activated archives. While very little of such information is consequential, it is always distracting.

We live in the Age of Distraction, where we are distracted away from knowledgeable insight into the conditions of our existence toward the pleasure of consuming inconsequential, mediated, and disposable information. The Age of Distraction makes entertainment out of information, making marketable the pleasure of processing data and the sense of satisfaction gotten from being informed. And while it may not be possible to control the flow of information—and, anyway, it would make no sense to try—there is a growing need to comprehend and exert some control over what distracts us. When information becomes entertainment, we are very likely to confuse distraction with paying attention. One result of such confusion is that we may lose the ability to control our modes of attending to the world around us. For example, the main agenda of the television industry is to convince its audience of the indispensability of its product, from newscasts through talk shows and from network prime time through cable programming. Sophisticated marketing techniques maintain the illusion of rapidly passing time, the necessity of continuous updates, revisions, and time-saving structures of delivery. Thanks to the ingenuity of television marketing, most Americans are convinced that they are very busy, with little time for leisure, and thus in great need of mass-media services.

A shared sense of having little time at our disposal is a consequence of being shaped by the Age of Distraction. With a sense of excess time, one might choose one's distractions with great care, setting aside time for activities selected for enrichment or cultivation as well as for mindlessness and oblivion, if desired. Conversely, with a sense of time's scarcity comes the willingness to abdicate the selection of distraction to services, from tour companies to television sets and Internet service providers. Certainly, companies that operate such services want us to see the information environment this way, and they profit on the shared sense of time's scarcity. However, the creation of time's scarcity is the result, not the cause, of the rise of such services. A longer view of the situation is illuminative. At the 1939 World's Fair, a *New York Times* reporter witnessed a demonstration of a prototype TV and was incredulous. "The problem with television," according to one of its earliest critics, "is that the people must sit and keep their eyes glued on the screen; the average American family hasn't time for it" (qtd. in Burstein and Kline 56). As we know, today the average family finds more time for TV watching than for any other activity. Television and the information environment have radically altered shared concepts of duration, so that an American in all seriousness can spend six hours in front of a television or PC screen and then testify to feeling the stress of time constraints.

Tremendous amounts of time are allocated to information processing because the experience afforded by it is so pleasurable, especially when the source of the information makes no cognitive demands on the viewer. The laugh track on television sitcoms provides a good illustration. If the television program were simply a broadcast comedy performance, there would be no need for it. However, on television information is provided not for analysis or even reaction but for consumption. We don't watch the sitcom to encounter humor or to engage in comic visions but to enjoy the pleasure of a superior position in which we can be understood as one who knows what is funny. This is why the laugh track, an objective absurdity, possesses its function on television. The danger of not getting the joke, among life's more humiliating experiences, is removed from television's information-processing experience. This service is generally emblematic of the information era, where we are continually assured that we have the latest and most complete news as it occurs and that our desires and needs are being met. The information environment is user-friendly and does not intend to burden its processing agents with ratiocination. Rick Altman reminds us that on television, "*anything really important will be cued by the sound track*" (42) and thus alert potentially distracted viewers to pay attention. When the laugh track sounds, the viewer is alerted to a situation in which information is to be processed as humor, which is, of course, the goal of the situation comedy. In general, television sound is employed to assert the place of television into the household flow, into teleconsciousness. "If we can be assured of being called back for the important moments, then it remains worthwhile to keep the set on even when we cannot remain within viewing distance" (Altman 43). As long as information sources are humming, the environment they maintain will continue to exert its influence on cognitive structures, thought processes, and our shared sense of time.

The satisfaction of consuming information parallels the consumer satisfaction of purchasing. In the exchange of money or credit for goods or services, the consumer experiences the satisfaction of direct interaction. However, because consumerism is not truly interactive—money, not experience, is exchanged—the satisfaction is fleeting and must be repeated to be sustained. Similarly, in the exchange of attention for information, the television viewer experiences the satisfaction of having gained knowledge through information. Again, however, because TV information rarely leads to knowledge—attention, not cognition, is exchanged—the satisfaction is fleeting and must be repeated to be sustained.

True, viewers can cultivate knowledge about television—about particular programs, for example—but this knowledge concerns a simulated world, like accumulating baseball trivia or reading fan magazines. The addictive qualities of television and shopping, like those of alcoholism and gambling, are based on the fact that nothing permanent is gained by an individual instance of the activity. After watching a television program, the viewer immediately wants to keep watching. Again, the contrast with reading is instructive. After reading *Moby-Dick,* the reader does not want to immediately start *War and Peace.* Reading literature of great substance usually gives one pause, and time is necessary for intellectual digestion. After viewing a television version of *Moby-Dick,* however, one is quite likely to flip over to the news or to whatever manages to distract while surfing channels. The two experiences are not equivalent or even comparable. When one reads *Moby-Dick,* one has a cognitive experience comparable to achieving a new perspective on life, a radical reorientation of sensibilities. When one watches the movie version, one is simply watching television and experiencing the pleasure of knowing something about a great novel.

The information environment was the pervasive, all-encompassing milieu of the late twentieth century. It doubtlessly will be the dominant structure of the twenty-first. Its promise fulfills the democratic ambitions of the nation toward universal entry and free, uninhibited exchange of views. Access to information, like access to roads and relocation, is a central tenet in the U.S. public philosophy. Human beings' information-processing abilities are accelerating as fast as did the ability to move or to travel as passengers in the first three-quarters of the century. We have come to understand the necessity of movement, and no one confuses a twice-daily commute of fifteen to twenty miles to the office with uprooting or escape. As a result, after fifty weeks of such travel, most Americans have a need to travel yet again on vacation getaways. Similarly, we must beware of confusing information that we consume for our entertainment with knowledge that is consequential. Just as the danger in habitual commuting is a distortion of one's sense of local distances, the danger in habitual information processing is the loss of the ability to distinguish infotainment from consequential knowledge. Entertainment distracts the mind by consuming time; in the form of information, it provides the pleasurable illusion of knowing. The result is the kind of passivity that flows from satisfaction, coupled with an arrogance that is produced by a continuous, structured reinforcement that one is in the know. Consequential knowledge, however, transforms the knower from one intellectual state to another and expands one's sense of now so that time is among the least of one's anxieties. Such knowledge does not satisfy, it agitates, so that the mind is enlivened with a sense of renewal and purpose. Consequential knowledge can be humbling, as one discovers not information but the depths of what one does not know and the expanse of what can be explored, absent the interference of consumption. We might be too distracted by the pressures of our busy lives to notice something as mundane as that phenomenon, unless, of course, we choose otherwise. For at base, what we choose to attend to is not a consumer but a political choice, affecting the course of history.

====="Don't Tread on Me"=====
Democracy and Distraction

A decent respect to the opinions of mankind requires that they should declare the causes which impel them to the separation.

I. To Pay Attention or Not to Pay Attention: That Is the (American) Question

Philip Gadsden, the South Carolina delegate to the Second Continental Congress, submitted an early design for an American flag to the Congress in 1776. Known today as the Gadsden flag, it depicts a coiled rattlesnake above the motto "Don't Tread on Me." The motto was a popular one during the revolutionary period, projecting well the image of an occupied, industrious people. The citizenry may seem inattentive, it may appear to be distracted by commerce, by private ambitions, and by its various pursuits of happiness, so that it seems unaware of dangers or challenges to its political welfare. However, these distractions are willed and subject to immediate cancellation once attention is drawn to an issue calling for unity, strength, and the "rattling" of the snake—or, the full attention of the democracy. Gadsden flag advocates argued that the rattlesnake was a good example of America's virtues. The animal was unique to the American continent. Alone, a single rattle is silent; when shaken with others, it produces its ominous sound. Furthermore, a rattlesnake does not attack unless provoked, although to disturb one from its solitude is a fatal error. The sentiments of the Gadsden flag are embodied in the

Bill of Rights, with its guarantees of privacy, and throughout the nation in various local town and city mottoes. This book was written in Cranston, Rhode Island. On the town trucks and public trash cans of Cranston, settled in 1636 and incorporated in 1754, is an attenuated version of "Don't Tread on Me." The Cranston city flag features the phrase *Dum Vigilo Curo*, which means, "While I Watch, I Care." A typical American attitude, the words suggest that while we may seem distracted by other pursuits, if called on to watch, to turn away from what we are doing to take note of an issue worthy of our attention, we shall engage. The rattlesnake thus warns against confusing distraction with powerlessness, lack of commitment, or apathy.

The level of attention demanded of citizens in a modern democracy is much more than most human beings can bear. To fully participate as a citizen in democratic government means attending to neighborhood, town, local or county, state, regional, and national processes and policies. Full participation also requires attending—physically attending—meetings of school boards, town and city councils, zoning commissions, and state and national organizations, along with subcommittees and project committees attached to and serving each of these various institutions. The average working American has never allotted the time to manage anywhere near this level of democratic participation. At best, Americans participate in their democracy by staying informed on issues to the extent necessary for them to vote with some confidence. This may require watching the television news or reading a daily newspaper. However, the percentage of registered voters who have gone to the polls in recent years suggests that even that level of attention has proved taxing to a growing plurality of Americans. Ironically but not surprisingly, as information sources proliferate, the capacity of people to attend to political issues declines. Because it seems as if there is so much to know, many react with mild forms of attention deficit, under the assumption that keeping up with events is simply impossible. Complicating, or aggravating this situation is Americans' traditional right—made emblematic by the coiled rattlesnake—not to pay attention unless provoked.

The greatest danger to a democracy, nonetheless, is an apathetic population that cannot master its powers of distraction and, as a result, becomes willing to wholly abdicate the exercise of sovereignty. The information environment poses a series of challenges to democracy. Its inherent sense of time's scarcity frames political action in such a way that almost any significant event or occurrence is perceived as a crisis and thus is accompanied by the illusion that we lack the time adequately to process its importance. All we have time for is the singularity of the event itself, fully disclosed as the media counts off Day 1, Day 2, and so on, before the next crisis occurs. Where time is experienced as scarce, deliberation is especially difficult because events seem relentlessly unprecedented. The plethora of intrusive, inconsequential information obscures the ability to assess significance, giving a cartoonlike aura to weighty developments and complex human motivations and subjecting all phenomena to the leveling tendency of mass media. At the same time, an attenuated sense of the now raises insignificant matters to unmerited levels of importance, relying on their capacity to garnish attention through titillation or novelty. The glut of information may also obscure the true significance of our own acts, as we come to see reality in accordance with conceptual frames provided by the digital age. Being called on repeatedly to process inconsequential information obscures the complexity of genuine events in human affairs and attenuates one's capacity to comprehend developments and occurrences with multiple explanations and nuanced significance. One thinks here of children who bring guns to schools and playgrounds and then surprise themselves at the consequences of deadly action. The environment in which contemporary citizens make decisions and formulate opinions on matters of community and state is fraught with phenomena and stimuli that threaten their ability to act responsibly. The environment cannot be changed, at least not readily. Nonetheless, one can master the terms under which one is compelled to think and to live.

The profit potential for those who can manipulate citizen attention is tremendous, and in many ways, consumer society depends on the power of advertisers and product (including information-product) designers to distract people into spending situations and long-term debt arrangements. The relationship between politics and profit making is as old as democracy itself, and no democracy has been entirely successful at keeping civic and business interests distinct. Democratic participation alone is seldom profitable, and so the attention of citizens and politicians is easily steered away from commonweal to consumption. The desire to consume, for example, buttresses the acceptance of longer working hours and allows fewer hours devoted to public affairs. If nothing else, consumerism has saved the work ethic. Gaining additional clients or working overtime will lead to much more satisfying methods of distraction than will working to understand changes in the distribution of wealth or the intricacies of the debate over capital punishment. Similarly, using one's position as a public official for economic gain is a powerful and consistent distraction from the duties of government. Even the free press falls victim to the greater distractions of sensationalism, and much of what has passed for news in recent decades is gossip, crime reports, and examples of the human grotesque. The circus and the midway have always been attention getters, and politics alone cannot compete with the compelling distractions of profit and consumption.

Citizens who engage in political action may contribute to the democracy but hardly contribute at all to the consumer economy—unless they are encouraged to see the two realms as continuous and thus participate with cash. Increasingly, the distinctions between the political and consumer marketplaces are indistinguishable. Political marketing companies and campaign management staffs know that citizens are readily distracted by their desire to consume goods and services; thus, packaging political ideas like consumer goods is an effective strategy. For example, political polling turns election issues into free-market bidding contests as a means of attracting voter attention. News agencies report on electoral contests as if they were horse races or competing market shares, making relative standing rather than issues or ideas the content of coverage. Those who attend to public matters, of course, must do so within the structures of corporate media. The polling company thus serves as auctioneer, barking the current voter-market status of those running for office, asking continuously for yes and no answers (bid or don't bid) to complex or multifaceted questions. Those placard candidate signs placed in political supporters' suburban front yards hardly do more than announce the securing of a household's vote: "This mortgaged property prefers this particular candidate." The similarity between the candidate placard and the real estate company's "for sale" sign emphasizes the correlation. If the market stays high for candidate X, he wins. A popular campaign slogan asks citizens whether they are "better off" now than at the last election, implying an economic and consumer-conscious gauge to political assessment. Seldom does the campaigner ask citizens to judge whether their fellows are better off or whether social justice has improved. As a result, in a prosperous democracy, a plurality of those who have the right to attend to political matters will not do so, preferring (by choice or seduction) alternative sources of distraction.

Historically, capturing U.S. citizens' attention for politics has taken some effort and much careful planning. The American political system has structured into its operations periodic calls to attention in the form of frequent elections, legislatures regularly pass "sunshine laws" to make governmental operations more accessible to the public, and the Constitution guarantees the operation of a free-market competition for attention in the form of a free and politically powerful press. Cable television provides coverage of local and national legislative bodies; government agencies regularly post (or dump) materials on web sites. However, because the mass media are owned and operated by corporate interests, some of which possess missions contrary to the interests of a democracy, citizens are both dependent on the press for information and justifiably wary of mass-media sources' encroachment into their cognitive liberties. The marriage of consumer and political culture in the United States means that disaffection in one realm will almost always encroach on the other and that neither realm may be comprehended in isolation. Consumers have rights, for example, largely because citizens possess the same, and those who do not vote are deadbeat citizens, like consumers who won't spend. An election thus closely resembles an "end of the lease" (or the term of office) sales event (twelve hours only!) and voter/consumer confidence is measured by the turnout or the election receipts as much as by the results. Under such circumstances there is no need to be concerned with selling out; marketing structures have long since captured the political imagination of the mass media and of most citizens.

In many ways, the information environment emerges organically from the political culture of *Don't Tread on Me* as it expands, toward infinity, the sense of privately accessed, individually consumed assets—assets, in this case, in the form of information sources. The Gadsden political realm is one where citizens prefer to be left to their private distractions, including, paradoxically, knowing a lot about what others are doing. In the Age of Distraction, then, citizens tend to judge the success of the political realm by the degree to which it leaves them alone (or provides the greatest degree of freedom to consume), and so it is no surprise that the intensely private distractions of the Information Age should capture Americans' attention. The Internet is no assault or departure from American ideologies of privacy but an expression of their most fundamental aspects. Internet access reifies the American mind as Emerson's transparent eyeball, the all-seeing orb with its original relation to the universe and, now, to all its data. Quite expectedly, users' initial impulse is acquisition, and consumer usage dominates research into web activity. Equally, among the first defenses of Internet access (in schools, for example) trumpets its potential to increase citizens' knowledge base, to provide greater access to information sources necessary for full democratic participation.

Consumer and democratic processes have evolved in tandem in the United States, each developing means of capturing constituents' attention and borrowing methods from the other, each stressing the centrality of freely made choice. At the same time, the nation's public philosophy contains elements that serve to structure the idea of politicized distraction into the national ideology, so that Americans are drawn away from private matters toward issues of state with periodic regularity. American citizenship, unlike citizenship tied to ethnicity, is based on state rituals not on religious or ethnic continuities. Eric Hobsbawm reminds us that the word *patriot* is rooted in the American experience and originally meant "those who showed the love of their country by wishing to renew it by reform or revolution. And the *patrie* to which their loyalty lay, was the opposite of an essential, pre-existing unit, but a nation created by *the political choice of its members* who, in doing so, broke with or at least demoted their former loyalties" (87; emphasis added). The idea of making a political choice is ritualized into the voting process and in the state celebrations that mark heroic decisions made in the past—Abraham Lincoln's, George Washington's, and Martin Luther King's as well as choices made by soldiers, laborers, and most celebrated of all, revolutionaries. To choose is to exercise complete control over competing areas of distraction and to experience, in attenuated but no less legitimate fashion, the essence of free action. On a continuum, voting for one candidate over another shares cognitive space with more heroic decisions, such as the decision to free slaves, to allow the transfer of power, or to enact a dream of justice. To vote is thus to exercise sovereignty over what one will attend to and to focus for that instant on the cognitive link between what distracts and what eventually, through democratic processes, will constitute reality.

"Choosing" constitutes a vital cognitive exercise in U.S. culture, whether choosing candidates, policies, or market products. Cognitively, choosing is an effort that counters distraction or controls it, demarcating the capacity of the mind to wrest control over what it will attend to. American civic life contains a number of ritualized political and ideological features, so that in the United States, public culture is far more likely to be based on or to possess major elements of state politics than to be rooted in religious or ethnic rituals. An ideology of freedom depends on making plain the connection between choice and eventuality, assuring that public decisions made today will affect the future in measurable ways. We tend to forget presidents who maintained the status quo (however pleasant it may be to live in such eras) while making heroes of those who made highly consequential choices.

Similarly, Memorial Day rhetoric consists in high measure of admiration and gratitude for those who choose to serve their country (even though most had no choice); Thanksgiving acknowledges the volition of settlers and natives, who choose cooperation (even though they eventually chose otherwise); and July 4, most notably of all, celebrates the choice of independence (a choice made by one-third of the population at best). The parenthetical comments in the previous sentence are not indications of cynicism but meant to stress the importance of framing issues in terms of choice to a distracted people. Even in cases where the existence of options is historically debatable, in the popular imagination, the element of choice is stressed. Richard Nixon, who was surely headed for impeachment, may come to be remembered with respectful admiration because, in stressful times, he choose to resign rather than extend the national crisis of his presidency. Within the context of democratic distractions, evidence of choosing what to attend to, what to think in the face of competing or contradictory demands, is evidence of a democratic process operating efficiently. And more, evidence of choosing proves mastery over distraction because choice displays the mind's capacity to decide, thus assuring the authority of the individual over the (democratic, informational, consumerist) environment.

Distraction properly controlled leads to renewal. Minds free for distraction will eventually return to their socially engaged focus of attention with renewed vigor, assuming that the distraction was not obstructed or preyed on by exploitative interests. Minds not in control of distractions will be indecisive, ambivalent, and ineffectually passive. The value of managed distraction maintains its place in American public thought because it may be traced to the founding of the nation itself, if not to the colonial era. The freedom to wander, intellectually and physically, is structured into the national ideology, into received cognitive processes, and into the way Americans think. Perhaps the most popular phrase in the dictionary of American ideological thought is the *pursuit of happiness*. The phrase sanctions distraction with the authority of the nation's founders and elevates distraction, within the structure of the Declaration of Independence, to a level of importance equivalent to that of human existence and political freedom. The declaration refers to "certain inalienable rights" and specifically enumerates three: "life, Liberty, and the pursuit of Happiness." The pursuit of happiness expands on the original, Lockean idea of the pursuit of property by incorporating into it a wider range of endeavors that may distract human beings and bring them pleasure. Governments are thus instituted, by the declaration's logic, to protect physical existence (life), political freedom (liberty), and cognitive distractions (the pursuit of happiness).

Americans, from their initial declaration of national statehood, have thus asserted the right to do things more or less for the hell of it—for happiness, pleasure, and to institutionalize the right to distraction. Critics who point to American citizens' hedonism and materialism neglect the importance of these pursuits to the nation's origins. Such qualities have made Americans particularly vulnerable to confidence games and swindlers and have produced a wide range of crass and vulgar behavior; nonetheless, these qualities of self-pursuit are not easily distinguished from larger definitions of American national traits. The right to privacy, embodied in the Fourth Amendment to the Constitution ("The right of the people to be secure in their persons, houses, papers, and effects") is a more positive way to phrase the right to distraction, the right to choose not pay attention to the state and the concomitant right not to be hassled for it. The sentiments of the Gadsden flag, "Don't Tread on Me," thus form the basis of rights to privacy and security, rights that are fundamental to any conception of U.S. public philosophy. Such freedoms have the effect of assuring a patriotic citizenry by linking a preference to be left alone with loyalty. Looking away may well be as vital to intellectual engagement as is extended concentration, and in this context, it is ideologically revealing that the pursuit of happiness be included among the revolutionary era's political claims. *Happiness* may well be among the more subjective, fleeting, and socially undefined terms employed in American documentary history. At the same time, the term is employed regularly in human efforts to describe one's state of mind. In the United States, the quality of one's distractions is a political matter, according to nation's foundational document, and one has a fundamental, state-endorsed right to pursue it. Political existence and independence in the declaration is thereby linked inextricably to life itself, to liberty, and to the right of cognitive distraction.

Conversely, we are often told that those who do not look away truly care about their fellow human beings and are engaged in human progress. Those who forgo or who have little time for aesthetic matters, for pursuits of happiness, are those who accomplish the aims of justice and progressive change. Intellectual battles have been waged between those who find distraction invigorating to the cause of human welfare and those who see distraction as detracting from what needs to be done. Activists who refuse to look away and who refuse to ignore social evils but will direct their gaze at what needs our attention are considered to actually move events forward. That is somewhat true, but it's not what I'm talking about at all. When we have looked long enough to know that this particular way of looking is not going to get us anywhere, that the problem or the crisis or the issue is insurmountable using the powers we have brought to bear, then it is time to look away and allow for the distraction of unthought possibilities. Conflicts between human minds more aesthetically grounded and those more grounded in materialism or activism are purely local. The species as a whole requires both modes of being. The declaration's inclusion of the wildly open phrase "*the pursuit of happiness*" within its very specific enumeration of rights and grievances introduces the political value of distraction into American ideological processes and intellectual traditions. The process of looking away and the political right to pursue happiness are directly related to rituals of renewal and release and to "revolutions" in the way citizens conduct themselves and fashion their democratic goals and procedures. Such rituals are structured into the very reason for the existence of the United States; they are what defines and justifies its independence.

The ability to manage one's distractions is absolutely crucial to the health and welfare of the democracy, and the act of pursuing happiness is fundamentally political. Those who cannot manage distraction, who allow their minds to exist at the complete mercy of market forces and consumer, mass-media seduction, pose a real danger to democracies because they have abdicated the *pursuit* of happiness in favor of the passive reception of happiness. Passive reception is apolitical at best; at its worst, it impedes progressive or necessary change in the social fabric. Those who fail to pursue happiness—which means, at the very least, consciousness of the relationship between one's cognitive processes and one's livelihood—weaken democratic systems by contributing no intellectual stimuli. The American democratic system is structured with the assumption that the citizenry is heavily distracted by its various pursuits; a major presupposition of U.S. public philosophy is that its people are cognizant of the competing demands on their conscious lives and mark their lives by responsibility for the arrangement of these demands. American democracy values distraction highly; it depends on individual pursuits for its very necessary standard of living and on periodic attention to public affairs for the maintenance of its political processes.

A number of cyclical or recurring political events are structured into American public life, under the assumption that regular calls to attention are necessary to the workings of the democracy. Every four years in the United States a new president—even if it is the same one as before—offers an inaugural address. *To inaugurate* means to confer solemnity or sanctity on something and, more importantly, to enter into a course of action formally or ceremoniously. The word is commonly used as a more formal synonym for words like *initiate, introduce,* or *begin.* Every four years, then, the American political system, marked by its presidential figurehead, creates the illusion of starting over again with a national ceremony of renewal to mark the occasion. Journalists keep track of the president's first hundred days in office, to see what kind of start is made, and it is almost as if history begins again, and we are there again, in the first days. Similarly, every year, on July 4, Americans celebrate the issuance of a document proclaiming the end of political ties to one nation and the commencement of independence. In ceremonies around the country, the Declaration of Independence is read—or at least referred to—as a means of thinking about the efforts that went into the nation's establishment. Strong efforts at cognitive direction are universally employed in the form of fireworks—a display that some may find confusing or link in error to military traditions but that is more accurately understood as a ritual of riveted attention. For a while, in the dark, everyone is looking at the same thing, communally distracted by the sounds and colors now associated with the nation's existence. Local equivalents and adaptations of these rituals are practiced as well, with state and community "birthdays" celebrated with various forms of distraction from parades to carnival events.

Historically, the place itself, the American landmass, was encountered as a geographical interruption. Intent on finding a quicker route to the east, Europeans became distracted by the American continents. In a perverse sense the colonization and subjugation of the area subsequently known as the United States was an extended, imperial distraction. Colonization projects were seldom the cornerstone of European foreign policy in the sixteenth century and by the eighteenth century had become a contributing cause of imperial decline in many European nations. Those who today wish to assign grand designs of imperialism to these centuries and identify a particularly horrific brand of Euro-sponsored oppression choose against emphasizing the historical fact that Europeans did not, from the beginning, set out to locate or inhabit another continent. Some left home looking for profit, some for religion, and some for adventure; most were thus so distracted by other imperatives that they did not see very clearly where they were or what effect they were having on the local environment or population. Throughout and long after the American Revolution, for example, most people identified themselves with their places of origin—as English, German, or Dutch colonists—or with their very local state or regional place within the United States. In his farewell address, George Washington instructed that the "name American . . . belongs to you in your national capacity," as opposed to (and meant to be more important than) "any appellation derived from local discriminations" (Boorstin 216). Even in 1796, the idea of America was a political and social distraction that the government sought to cultivate among its people.

The colonization of the American hemisphere by Europeans may well be less an example of historic oppression and more an illustration of what human beings are able to accomplish when they pursue their distractions. If the typical British colonist in New England could be born, live, and die in Massachusetts while maintaining a sense of herself as English, all the while encroaching on lands occupied for centuries by a native people of whose language, culture, and history she knew and wanted to know nothing, then we can begin to see the power of distraction to shape world affairs. For example, to see the settlers on whose sense of reality native populations could not and did not seriously impinge as conscious imperialists is to learn little of value from the actions of individual human minds within their specific historical circumstances. Seldom in human history have regimes thrived whose consciously acknowledged goal was to be cruel or inhuman. Rather, inhumane practices have been the byproduct of rational—and often thought to be quite compassionate—systematic pursuits. Rather than making villains of members of our species who have preceded us, we are wiser to ask what practices we perpetuate. What distracts us today from seeing the damage we are doing to ourselves, our environment, or the certainty of our future? When we say, "Don't tread on me," on whom do we tread unconsciously?

II. Captive Attention; or, Why Distraction Is Inherently Revisionist and Incomplete

One of the earliest documents in colonial American history is Mary Rowlandson's *Narrative of Captivity*, an account of her 1676 abduction by Narragansett Indians. Published originally as *The Sovereignty & Goodness of God, Together, with the Faithful-*

ness of his Promises Displayed, the text was intended as a testimonial to Rowlandson's ability to maintain her faith in God while suffering the trials of captivity. As a member of a Christian community, she focuses primarily on the extent to which her life is consistent with the precepts and prescriptive guidance of Holy Scripture. Most other stimuli, including the actions of the native peoples she encounters and the physical trials of her captivity, she treats as distractions from that mission. The Narragansett people are clearly identified as demonic, inhuman creatures. Here is a typical passage from her narrative: "God was with me in a wonderful manner, carrying me along and bearing up my spirit that it did not quite fail. *One of the Indians carried my poor wounded babe upon a horse;* it went moaning all along, "I shall die, I shall die." I went on foot after it with sorrow that cannot be expressed. At length I took it off the horse and carried it in my arms till my strength failed, and I fell down with it. *Then they set me upon a horse with my wounded child in my lap.* And there being no furniture [saddle] upon the horse's back, as we were going down a steep hill, we both fell over the horse's head, at which they like inhuman creatures laughed and rejoiced to see it, though I thought we should there have ended our days, as overcome with so many difficulties. But the Lord renewed my strength still and carried me along that I might see more of His power" (Vaughan and Clark 37; emphasis added). Later on, an Indian man joins her on the unfurnished horse with her child, presumably to keep her from falling off.

It is difficult to attend to passages such as these today and not conclude that they show evidence not so much of God's goodness (though they may) but specifically of the kindness of the Narragansett Indians. The material evidence no longer shows that Rowlandson's strength is renewed by God (to enable her to see more of God's power) but that her strength was renewed because the Indians allowed her to ride on horseback while many of them had to travel on foot. I've italicized portions of her testimony to highlight this evidence, which she, in her own analysis of the events, mostly ignores. While Rowlandson would see the Narragansetts as instruments of Divine Providence, we are inclined, in the present era, to grant them full subjectivity. They seem to find it tremendously amusing that a woman in the seventeenth-century wilderness—especially one who has chosen to plant her life in that wilderness—is unable to ride without saddle furniture, for example. Still, there were not enough horses for everyone to ride, and Rowlandson was allowed the luxury and the honor of not traveling on foot. At first, only her child is carried, but once the anguish of separation is clearly unbearable to the child, the Narragansetts allow them both on the horse's back. When her inept riding skills are displayed, a chaperone is supplied to steady her and her child on the horse.

The *Narrative* is sprinkled with examples of such acts of kindness. "In this travel because of my wound I was somewhat favored in my load," she announces at one point, although as we would expect, Rowlandson acknowledges God's intervention rather than the kindness of the natives for the lighter burden. Similarly, when a river is crossed, "some brush which they had laid upon the raft to sit upon" kept her injured foot from getting wet, "which cannot but be acknowledged as a favor of God to my weakened body" (43). Throughout Rowlandson's narrative, express and repeated descriptions of the Narragansetts as "inhuman" and "hellish" are undercut by evidence, apparent today but clearly unintended and perhaps unreadable in Rowlandson's era, of humane native behavior. The narrative does not intend but makes very clear the kindness and humanity of her captors. Readers today see this phenomenon by managing their powers of attentiveness over the text and becoming sensitive not only to what captured Rowlandson's body but to what both captured and escaped her attention. What distracted Rowlandson ("at which they like inhuman creatures laughed and rejoiced to see it") is thus centered by a modern consciousness, intent on seeing what those before it had ignored. We seize on the laughter of the Narragansetts to transcend Rowlandson's sense of reality and to find a sense of reality more aligned to present sensibilities, one that makes sense to us. What was at the center of Rowlandson's attention ("and carried me along that I might see more of His power") is more likely to be considered today as evidence of an ideology that distracted Rowlandson from what was really going on in the Massachusetts wilderness in 1676.

Rowlandson's book was published by those who had designed it as a testimony of faith. While it may still be read as such, the *Narrative* is more likely to be considered today as exemplary of a host of things Rowlandson either had no knowledge of or valued only slightly if at all. The felt superiority of historical perspective can be deceiving, however. Rowlandson's text works hard to instruct readers about the strength of her faith and about her constancy in the face of trial and suffering. The fact that we find her testimony less valuable for its expression of Christianity than for the clues it holds to English attitudes toward Narragansetts, colonial history, or Narragansett customs does not necessarily mean that we know something Rowlandson did not know. It does indicate a shift in what we attend to, away from the spiritual and toward the material, away from exclusive concern for those whose history and language we share toward those defined as others whom we now judge to have been historically overlooked, slighted, or mistreated. More importantly, though, our opinion of her experience indicates that whatever perspective we enjoy today (including modern relativism and the postmodern refusal to be implicated in any perspective) will in all likelihood be found quite limited in the future. Rowlandson was nothing if not secure that her view of the world was proper, advanced, and unimpeachable. In her mind, God's will was of greater presence than that of her Narragansett captors. And in her world, in the way her society was ordered, she was right.

Documents, texts, and written materials cannot have eternally fixed meanings and are able to maintain their significance—even their profundity—long after their original meaning has been lost in large part because of the inherent indeterminacy of language. This truism of contemporary literary study, the indeterminacy of the text, is fundamental to any understanding of U.S. political culture. One way significance is maintained is through the way in which language effects or provides structures for cognition. A person raised on the "pursuit of happiness" will come to recognize the significance of the phrase to patterns of thought and desire largely because it was learned before its meaning was realized. Or, in William Faulkner's well-known formulation, "memory believes before knowing remembers" (Faulkner, *Light* 119). Ideologies operate in this way, informing the structures of thought long before those structures are filled with content, with something known. U.S. public education is based on this premise, including strong component measures of civic instruction, commencing with the daily recitation of the Pledge of Allegiance. However, while the significance of words may change, as long as whatever meaning is agreed on is rooted in the same texts, a sense of national continuity will be sustained even in periods of revision and reform. Communities do not have to agree on interpretations, but they do need to be looking at the same thing.

Textual continuity in the United States works to alleviate the sense that a nation of such varied identities and origins, and a nation where the population prefers to be let alone, cannot possess cohesiveness. It is no surprise that the founders would reject the Gadsden design for the national flag, as "Don't Tread on Me" hardly invokes a sense of a cohesive nation to one unfamiliar with the community that has managed to evolve from that basis. By the evidence itself, the nation's ethnic, racial, and religious diversity cannot possibly maintain cohesiveness. Lawrence Levine has argued that "the United States harbored the kernels of division from the beginning and contained a multifarious population distinguished by race, religion, country of origin, philosophy, accent, language, class, and region" (119). Division has been endemic to U.S. history, and there has been no era in that history when serious lines of contention did not threaten its cohesion—from sectional antipathies though contemporary linguistic and racial divisiveness. The ideological imperative to pursue happiness means, quite literally, that public philosophy fully sanctions noncommunal and antisocial pursuits. Citizens are expected to go their own way. Immigration policies, beckoning others in the world to join the experiment in democracy, threatened U.S. unity by introducing yet additional stresses to its population. Levine thus advocates that we accept contention—it is the reason for practicing democracy, after all—and avoid seeing in threats to unity threats to the national mission: "The very strength of the United States—its size and diversity—inhibits if it does not make impossible the kind of conventional wholeness and stability that some of us seem to long for and even to invent historically so that the present is made to seem aberrant, and even culpable, for harboring the seeds of separatism and alienation when in fact those seeds have been present and have born fruit throughout our history" (119). The United States is a democracy *because* it is a contentious society, not despite that fact. If it were not contentious, there would be no reason to guarantee freedoms or establish procedures and limitations for the exercise of power.

The nation practices its public contention on the stability of texts whose words are written down and so do not change but whose meanings evolve and thus promise endless possibility. Much of what we know in the United States as political struggle, therefore, is in reality a struggle for the contemporary definition of historic terms—*justice, freedom, equality*—where significance is held as something distinct from meaning. We know these words are significant; we simply do not always agree on what they now mean in practice. Ideologically, significance is agreed on before meaning is established. It's best not to pay too much attention unless one's private and local distractions are tread on, unless redefinition or clarification is necessary to accomplish one's purposes. Most U.S. reform movements have concerned themselves with the expansion of the definitions of words and phrases already agreed on in the national vocabulary, such as the right to vote or the freedom to occupy public space or go to public school.

Paradoxically, distraction makes U.S. cohesion possible, the distraction born of the ideology *Don't Tread on Me*. Distraction in the face of contention manifests itself as indifference. Only through indifference to "race, religion, country of origin, philosophy, accent, language, class, and region" is U.S. civilization possible. Precisely because Americans insist that they be left alone, that they be granted and protected in their private distractions, the nation manages to survive while possessing characteristics that would (and do) threaten the stability of most nations with perhaps a less distracted citizenry. J. Hector St. John de Crèvecoeur's observation that in America "individuals of all nations are melted into a new race of men, whose labours and posterity will one day cause great changes in the world" (70) was as prophetic in 1782 as it is troubling to some today. What of difference? What of the integrity of cultures? Crèvecoeur predicted that difference would be supplanted by indifference, particularly regarding what was the most divisive social distinction of his time, religious sectarianism. In America "all sects are mixed, as well as all nations; thus religious indifference is imperceptibly disseminated from one end of the continent to the other, which is at present one of the strongest characteristics of the Americans. Where this will reach no one can tell," Crèvecoeur muses; "perhaps it will lead to a vacuum fit to receive other systems" (76). Religious and ethnic indifference continue to mark U.S. culture, and political movements for inclusion—most recently, multiculturalism—must work to revise and expand the significance of established political texts to include new constituencies. For two centuries, for example, the words of the Declaration of Independence have been used to justify the political rights of peoples never intended for inclusion by the authors of that document. U.S. political struggle often centers less on the integrity of peoples and more on the applicability and definitions of a priori political phrases and concepts to those peoples.

The fact that the meaning of words may change, that what was once overlooked or unrealized may subsequently distract and command attention, ensures that each succeeding generation will not only look on the world as if for the first time but in actuality, linguistically, for the first time confront its textual heritage. For human events to have meaning, they must be rendered into language. For language to communicate over time, it must be written down. But putting perceptions into words relinquishes control over the meaning of those perceptions, taking a great measure of the power to control meaning away from the writer and granting it to the reader or the listener. A succession of literary theorists have dealt with this phenomenon, beginning in the United States with New Criticism's notion of the intentional fallacy, or the denial that the author controls the meaning of the text. Various critical methods have provided literary scholars with tools to create and manipulate textual meaning—feminist, poststructuralist, historicist, psychoanalytic, reader-response—so that in literary studies today, the notion that a text has fixed signification set by its author has been thoroughly discredited. The idea of a readerly world, where meaning is not held and dispensed (and then discovered) but is actively created by the living generation, is fundamental to a democracy. It is also an idea a distracted citizen will find both comforting and taxing, for while it allows for periods of prolonged indifference, it also may demand a level of attention, in times of crisis, that may interfere with other pursuits. Nonetheless, if citizens are not empowered to move their world somewhere, there is no reason for the perpetuation of democratic processes.

Mary Rowlandson would certainly be appalled to learn that her religious message, so insistently stated and emphasized throughout her text, is so readily dismissed today as relatively, if not fundamentally, insignificant. However, she surely dismissed the interpretation of some interest—Christians in Great Britain, perhaps—as lacking in authority. Once rendered into words, the experience of the author, fictional or testimonial, is no longer the writer's property. Words fly from their context like emigrants, altogether accommodating themselves to renewed significance. Long after the original meaning and the contextual setting of human declaration have passed from memory, words, when read, live on, spawning new generations of significance. Influence, in such a linguistic environment, accrues to those who can read and interpret and convince subsequent readers of the veracity of their interpretations. In a democratic society, where government must compete with private interests, corporations, and nonprofit organizations to assign meaning to contemporary and historical events, literacy and interpretive powers are vital to the exercise of citizenship. Actions may speak louder than words, but words, in their silence, speak eternally.

In the middle of the dedication of the Gettysburg battlefield, in one of the nation's most important speeches, Abraham Lincoln said, "The world will little note, nor long remember what we say here, but it can never forget what they did here." The irony in Lincoln's words is that long after most Americans have forgotten exactly what happened at Gettysburg in 1863, they can recognize the opening phrases, "Fourscore and seven years ago, our fathers brought forth, on this continent, a new nation," and the closing, "of the people, by the people, for the people." The world has indeed remembered what Lincoln said on that battlefield, at least as well, if not more vividly, than what the soldiers did there. Acts that Lincoln suggested were impossible to accomplish with words ("we can not dedicate—we can not consecrate—we can not hallow—this ground") were in fact possible only by narrative declaration, only by capturing listeners' and readers' attention. Without words, without the assertion of the significance of acts taken by human beings ("The brave men, living and dead, who struggled here"), such action, no matter how heroic or singular, is certain to be forgotten. Lincoln's declaration of the impotence of words only added to the humility the address evoked and in turn contributed to the consecration of those who died on the battlefield. His speech focused attention on heroic death, and the eloquence of his words assured that the idea of national sacrifice would endure long after the passing of the particular events. As a result, the sense of urgency he sought to associate with the "great task remaining before us" is enhanced in the speech by the humble insistence on the insignificance of words. For only through the power of words can people be convinced that what is needed now is attentiveness and physical commitment to a cause. Lincoln seemed to know this fact, too, and he articulated an American article of faith when he confidently asserted, in the dedication of the battlefield, "It is altogether fitting and proper that we should do this" (Boorstin 436–37)

The United States is a textual nation, where national celebrations turn to founding documents and acts of public declaration for validation. Long after whatever national heroes did is forgotten, what they said ("I regret I have but one life to give for my country"; "Remember the Alamo"; "The only thing we have to fear is fear itself") remains as part of the national consciousness. The counterpart to American distraction is thus the compulsion to declare: to attract the attention of others, to turn deeds into words not simply for the historical record but for public dissemination, to be heard and read by others as a ritualistic act of community. The compulsion to declare, to garnish attention for public purposes, defines American nationalism as profoundly as any prior or resulting public policy or political institution. Lincoln's address at Gettysburg simultaneously invoked the tradition of declaration while conjuring the specter of its denial, which would mean the end of the nation. At the beginning of the speech, Lincoln evoked the nation's founding declaration by referring to the bringing forth "on this continent, a new nation . . . dedicated to the proposition that all men are created equal." He then denied the power of human declaration by evoking the horror that the world will "not remember" what we say here—and thus by implication, the political horror that the founding declaration be forgotten. For if the world forgets what we say, and if American declarations are ignored, the idea of the United States of America would become a historical impossibility, its cohesion quite literally rendered mute. American declarations must be heard and read and read and heard again. A distracted people know it intuitively and practice the arts of public avowal and declaration whenever tread on. Public textuality rather than ethnicity or revelation most solidly forges the nation's political bands, because only declaration arrests attention. The idea of the nation and its people's historical continuity is rooted in a faith in the endurance of the human voice and the written word. And that faith is manifested by contemporary acts of interpretation.

Among the first documents written by British colonists on the American continent is the Mayflower Compact, actually written onboard ship prior to making landfall at Cape Cod in 1620. The compact establishes a civil body politic among the signatories and enters into the historical record the fact that they have made their colonial voyage "for the glory of God and advancement of the Christian faith, and the honor of our king and country" (Boorstin 21). The immediate audience is God and the pilgrims themselves and the intent is to declare purposes and causes and to reaffirm the existence of their community—prior to its establishment. The compact thus focuses its signers' attention on their journey's purpose. The document's authors knew that the non-Puritans on board, known as strangers, were not invested in the community's mission, and so this declaration was a way to focus every pilgrim's attention on the same set of ideas and purposes. Recording their intentions on paper and then signing the document provides a foundational instance of a declarative, textually based community on the continent. Significantly, the document anticipates a distracted population. The Pilgrim separatists were already a textual, biblical people, and they sought to establish a Christian body politic whose source of law and social order would be grounded in the texts of Scripture.

The Mayflower Compact contains a narrative, causal declaration in its first sentence. The Pilgrims, after thanking God and identifying themselves as subjects of the king, assert that "Haveing undertaken . . . a voyage," they therefore "doe . . . covenant & combine . . . togeather into a civill body politick" (Boorstin 21). The body politic is predicated on the journey. *Having undertaken* the voyage enacts the covenant. The actual voyage had become misdirected, nautically distracted, and instead of landing at the mouth of the Hudson River, jurisdiction of the Virginia Company of London, the Mayflower landed in present-day Provincetown Harbor, property of the Northern Virginia Company. As such, the voyagers' initial patent, drawn at origins, was invalid. Nonetheless, historical events are often secondary to the structure assigned them by narratives of intent. The particular setting of the *Mayflower*'s landfall was secondary to the purpose for which the journey was undertaken. Attention was thus riveted on the movable community—that is, attention was drawn to the spiritual rather than the physical destination. While the separatists were not where they intended to be, they nonetheless had arrived and thus had a tale to tell and a document to sign. *Having undertaken a voyage,* all the people on the *Mayflower,* Pilgrim and stranger alike, form a single community, no matter where the ship lands, as long as the passengers remain capable of declaring themselves a people. "And by vertue" of this arrival, the compact declares that a community is formed: "And by vertue hearof to enacte, constitute, and frame shuch just and equal lawes" necessary for "the generall good of the Colonie" (Boorstin 21). The singular experience referred to in the Mayflower Compact is the voyage, and the simple narrative penned in its wake affirms the force of the imaginative process on the continent.

The Puritan agreement had little influence in the eighteenth century. Its importance became pronounced after American independence, however, when political leaders looked to historical documents for precedents and for sources of national cohesiveness. In 1802 John Quincy Adams suggested that it was "perhaps the only instance, in human history, of that positive, original social compact, which speculative philosophers have imagined as the only legitimate source of government" (qtd. in Boorstin 24). Throughout the nineteenth and twentieth centuries, the Mayflower Compact would be cited as evidence of an American tradition of self-government, despite the fact that democracy was far from the minds of those who signed the original document. Significance, again, both precedes and outlasts meaning. Since the establishment of the United States as an independent nation, it has been clear that the Mayflower Compact possesses significance; what has evolved is the meaning of the text itself. The signatories do perform an act of foundation, and they do intend a government, and no matter how often we remind ourselves about theocracy and authoritarianism in Puritan New England, the usefulness of the Mayflower Compact in extending back in time a tradition of American declarative textuality prevails over the accuracy of origination. The compact 's words do not belong any longer to Puritan authors but to the succession of contemporary readers who have found its text useful. If the document did not serve present concerns, its name would be forgotten. The document portends the politics of a distracted people, speaking into being a community in the face not of resistance as much as inattentiveness. By arresting the Word, as the Puritans would know, the act would follow.

III. DECLARING ATTENTION IN THE MIDST OF DISTRACTION

For the United States on the North American continent, the beginning is truly marked by the Word. It is of great significance to American ideological history that the nation celebrates its independence on July 4. The Continental Congress passed a formal resolution of political independence on July 2, 1776, two days before finalizing the language of the public declaration. July 2 is thus the historically accurate, technically precise, and legally binding date of American independence. On July 2 the Continental Congress approved the resolution brought forth by the Virginia delegation, led by Richard Henry Lee: "These United Colonies are, and of right ought to be, free and independent States, that they are absolved from all allegiance to the British Crown, and that all political connection between them and the State of Great Britain is, and ought to be, totally dissolved" (Gerber 20–21). With the passage of this resolution, the deed was done, the act of revolution, or of treason, accomplished. Nonetheless, Americans celebrate what was eventually said over the actions taken previously, marking independence with its declaration and not by its resolution. And again, Lincoln's assertion to the contrary, long after Lee's resolution fades to a footnote in the historical record, the subsequent declaration of causes and principles has entered the national fabric.

The Declaration of Independence is fully cognizant of the fact that it follows a previous resolution when it affirms that the colonies "are, and of Right ought to be Free and Independent States." The formal declaration is thus a consecration of sorts, of the same impulse as Lincoln's, except that people were about to die rather than lie dead already for the words spoken. The colonies were already free and already resolutely independent, according to the July 4 document; what was lacking was a freely enacted declaration, the performance of independence. Furthermore, it was crucial to distract colonists from their private business and gain their attention for the cause of independence. The historical record suggests that as much as one-third of all colonists were neutral in the revolution, while another third were Tory and another third favored the rebellion. For such a challenge, a simple act or resolution was not enough. Who will die for redundancies? The declaration thus asks Americans to stand not so much for independence (because they already possessed that quality) but for their right to declare their legal status, to perform their independence before an audience of the world, for "the opinions of mankind," as it was stated. "Lives . . . Fortunes, and . . . sacred Honor" were thus pledged by the signatories not specifically to self-evident truths or to revolution but to one another and "for the support of this Declaration" (Boorstin 89). How often have we heard an American say that she disagrees with a statement but supports the right of the speaker to declare his opinion?

A recent study of language and literary form in colonial America refers to the "improbable claim that the United States was actually 'spoken into being,'" referring thus to "the American sense of national fabrication as an intentional act of linguistic creation, the belief that the nation was made out of words" (Looby 4). The claim is improbable but not incredible. The nation certainly was made of more than words—lives were expended, battles fought, treaties negotiated—but none of these acts would achieve significance as community props without their transference into texts to be read and remembered. Turning acts into words is by no means peculiarly American; it is as old as human speech. Nonetheless, in the American experience we often see acts and words as simultaneous endeavors, so that the act of declaring becomes a vital third category of human enterprise in American history. There are acts (battles fought, resolutions passed), there are words (novels, poems, lyrics), and there are acts of declaration, or performatives, utterances that are also properly considered as deeds, utterances through which, in being spoken, the speaker performs a particular act. Americans celebrate the declaration and not the resolution of their independence because the act in itself is somehow understood as insufficient and is likely to be forgotten. A text of declaration, however, distracts, arrests attention, and becomes part of consciousness itself. In a nation where the people's ideology is "Don't Tread on Me," these declarations are the footprints of those exceptional instances when a distracted population was aroused to attend to public affairs.

The temporal and declarative origins of the United States attracted the attention of the French philosopher Jacques Derrida, who made it the subject of remarks made at the University of Virginia in 1976. "But this people does not exist," Derrida declared. "They do *not* exist as an entity, it does *not* exist, *before* this declaration, not *as such*." The performative nature of the foundational document is thus its most striking attribute. "If it gives birth to itself, as free and independent subject, as possible signer, this can hold only in the act of the signature, a sort of fabulous retroactivity" (10). And more: the ur-American text simultaneously eliminates one existence, when it dissolves ties to England, and gives birth, by signed declaration, to another. This liminal state is the site of historic American origins and its perpetual state of being: between the end and the beginning, between one meaning and another, the state of declaration, when something is about to distract us from one mode of attentiveness to another. American optimism is born of this moment, the expansive now, the state of possibility that arises from perpetual contingency. The moment does not pass in American culture but hangs and is re-created in various public rituals of renewal, rebirth, and new beginnings. In this way the world's oldest democracy maintains the sense of itself as the world's youngest nation. A declaration is something spoken, it exists only in the successive incarnations of the present; when written down, it becomes subject to successive interpretive acts. U.S. public culture is born of this sense of time's abundance, which may explain why its citizens have reacted with anxiety to contemporary arguments for time's scarcity. Much attention has gone to study the spatial existence of the United States—its expansive frontier, the policy of manifest destiny, its global influence. But as much as it depends on geography, the United States is equally rooted in a particular sense of time. We know precisely when the nation began, and public philosophy has expanded that moment to ritualistic proportions, the recurring now of celebratory acts.

We ought not to forget that "the Declaration was written to be read aloud. . . . On July 4 the Continental Congress ordered not only that 'the declaration be authenticated and printed,' but that 'copies of the declaration be sent to the several assemblies, conventions, and committees, or councils of safety, and to the several commanding officers [so] that it be proclaimed in each of the United States'" (Fliegelman 25). The primary audience for the declaration, despite its appeal to the "opinions of mankind," was American citizens—or, Americans not as colonials but as common members of the human race, as representatives of everyone. In English law, the tradition of the declaration served to cast public matters in "plain and certain" language (Maier 149) so that the contents could be readily understood by anyone, including especially the common person not likely to have patience with the obtuse abstractions of political philosophy. The assumption behind the tradition is that if materials issued by formal declaration would have an immediate impact on the lives of hearers and readers, "a decent respect to the opinions of mankind" necessitated that the materials be comprehensible and public. The American document announced political independence, evoked the cause of nationhood, and inspired soldiers on the battlefield—all in the common language of the day. It also preserved certain linguistic phrases that would become a part of American language for centuries. Pauline Maier argues that in the declaration, "Americans had to do more than demonstrate that the British Crown had forced them to the measure. They needed to overcome fear and the sense of loss, to link their cause with a purpose beyond survival alone, to raise the vision of a better future so compelling that in its name men would sacrifice even life itself" (95–96). The gravity of the situation is thus articulated and cast in the language of necessity: "When in the course of human events, it becomes necessary . . ." The cause is thus enacted and performed, from the founders' imagination into the sense of the era, into the minds of an audience sufficiently distracted by a plain and convincing argument that what was occurring on this continent at that time was of great and enduring significance.

Americans in the revolutionary generation were entering frightening political ground on which the future would become wholly uncertain. In subsequent imaginations, the nation envisioned by the Declaration of Independence would be linked, for the purpose of forging political continuity, to the covenant enacted aboard the immigrant ship the *Mayflower*. Both documents meet the tenuous uncertainty of desired action with the confidence of declarative performance designed to rivet attention. Authorities aboard the *Mayflower* feared that by landing in a place different from that provided by their charter, they would lose control over their passengers, who would not be bound to them in Cape Cod. That fear was alleviated by the declaration of a civil body politic, entered into by all aboard ship. Similarly, the political, economic, and physical risks implied by Lee's resolution of independence were certain to strike fear into the hearts of American colonists. In the vacuum created by the resolution of independence and to compensate for the profound sense of loss that surely accompanied the rejection of former political bands, the declaration enacted and promised an existence far better and of much greater significance than the one being rejected. The declaration performed a place where all men were created equal and where all were free to live as they wished, in pursuit of that which they found pleasurable. The colonies were already established on the principle of improvement—why migrate if not for betterment? The declaration elevated that principle into an ideological sensibility and bequeathed to posterity its assumption that such present personal desire ("the pursuit of happiness") is an unalienable political and human right. What people were doing now, in whatever pursuit they followed, would thenceforward be understood as not only a personal but also a profoundly political act. The new nation, in other words, guaranteed the right of distraction and politicized the nature of attentiveness.

The audience assumed by the declaration is deeply distracted and must be roused from its private pursuits toward political action—political action engaged in to guarantee private pursuits. In a May 8, 1825, letter to Henry Lee, Thomas Jefferson explained that in his draft of the document he sought "neither . . . originality of principle or sentiment, nor yet copies from any particular and previous writing." Rather, the document "was intended to be a expression of the American mind, and to give to that expression the proper tone and spirit called for by the occasion. All its authority rests then on the harmonizing sentiments of the day, whether expressed in conversation, in letters, printed essays, or in the elementary books of the public right, as Aristotle, Cicero, Locke, Sidney, etc." (Koch and Peden 719). A distracted public would thus recognize in the document ideas and phrases drawn from ancient and recent authorities and acknowledge that the "course of events" had made this action inevitable. The appeal to the opinions of mankind was an appeal to common sense, an act of justification in which the thoughts of the Continental Congress are shown to be in keeping with the accepted wisdom of the era. Breaking political bands is an act of revolution. The Declaration of Independence redefined this particular revolution away from willful self-aggrandizement and toward the universally accepted concepts of historical necessity and the laws of nature—and in the local context of the colonies, toward the preference to be left alone. Paradoxically, the declaration is a call to political action in the name of the right to attend not to political matters but to private pursuits and individually defined distraction. The snake had been tread on, its principles of privacy violated.

Importantly, the document's language assured that the causes outlined would not be misconstrued as selfish or base but universally just and, by implication, exportable. The American Revolution would be construed as idiosyncratic and thus less worthy of support, if interpreted as a local act performed by men distracted from loyalty by profit or self-aggrandizement. On the contrary, the declaration portrays a population that would prefer to be let alone to go about its business than be drawn into political action. A sanctioned distraction may be one of principle, a selfless act, even a cause to die for; an unsanctioned distraction is a dereliction of duty. The distinction is drawn with care, and attention to more than simple political independence and personal gain is thus declared. There is at once a sense of intellectual distancing and a series of gestures toward universalizing in the document, an assertion that what distracts Americans at this juncture ought to be of interest to all human beings. The signatories to the declaration assert an ability to see their historical situation from a perspective much wider and more inclusive than that of simple British or American interests. As such, the declaration "implies the possibility of *intellectual* detachment or independence, which means that our thoughts concerning justice, for example, are not wholly determined by non-rational or sociological forces" but have origins in universally recognized principles. As such, Paul Eidelberg suggests that the assertion of freedom from baser motives provided solid ground from which to project a national existence, a cause that must by necessity be rooted in a sense of difference and superiority. "In other words, the statesmen of the Declaration affirm the possibility of moral insight, which presupposes the freedom of the intellect from external compulsion" (Eidelberg 13; emphasis added). The document's opening words are therefore crucial, as they immediately establish a universal perspective on the contention between Britain and the colonies, by means of its omniscient viewpoint. The controlling perspective of the "Course of human events" simultaneously contextualizes the struggle within eternal standards of judgment and sets up a series of obligations that a decent people will be compelled to meet (first, to declare the causes that impel them) so that they may acknowledge cognizance of their place in history by issuing a statement for others to consider and to judge.

The intellectual detachment declared in 1776 also provided one basis for the sense of American exceptionalism that runs throughout the nation's history and historiography. This notion includes the sense that what happens in the United States is seldom limited to the local or national but possesses global significance because the United States is not *a* nation but *all* nations, all humanity. The idea of

American exceptionalism is fueled, furthermore, by the abstractions on which the United States declared itself into existence—the detached, moralistic, and highly indeterminate content of its revolution. Contemplating the significance of these matters has occupied American studies scholars for decades. Recently, in a book titled *American Exceptionalism,* Seymour Martin Lipset argues that "America continues to be qualitatively different. To reiterate, exceptionalism is a two-edged phenomenon; it does not mean better. This country is an outlier" (26). Not superior, necessarily, but "outlier"—an exception to the rules governing other nations because of its perspective on itself and on the world and because of an ideology rooted in moral principle rather than ethnicity or political policy. Lipset suggests, for example, that the "same moralistic factors which make for patriotism help to produce opposition to war." Similarly, a concern for civil liberties leads Americans to oppose gun-control laws as vehemently as they object the application of crime-control measures. A strong sense of individualism "both weakens social control mechanisms, which rely on strong ties to groups, and facilitates diverse forms of deviant behavior" (290). In all these examples, allegiance to a principle, rooted in a text somewhere ("I pledge allegiance"; "equal protection before the law"; "all men are created equal"), results in wildly divergent behavior, making life in the United States seem incomprehensibly contradictory to outsiders.

The solution to the enigma lies in the history of exceptionalism, as Lipset asserts, or in the tradition of intellectual detachment born of the indeterminacy of textuality. Significance both precedes and survives meaning. Significance occupies this position, subject to eternal revision, because of the expansive now in which Americans have historically operated. There has, simply, always been time for the rethinking and revising of core concepts necessary to maintain the present generation's freedom to rework the past's legacy. Because Americans base their national cohesion on abstractions open to the vagaries of interpretation, and because they are schooled very early in a tolerance for divergent readings ("I will defend your right to say that"), they can quite comfortably live within coexisting inconsistencies. We teach our children, in school every day, to think of their nation as simultaneously indivisible and tolerant of liberty and justice for all. The cohesive incompatibility of the two values is inherent in the ideology being inculcated. There is time to work these matters out in the expansive now of our existence. Everyone pursues his or her private distractions and is treated equally and the nation remains whole as long as the common denominator of public existence is suspension of completion. The politics of personal pursuits merges imperceptibly into a public morality.

The importance of moral grounding to the language of the declaration cannot be overestimated. To make legitimate claims to a distinct and independent political entity as well the right to be left alone to their private distractions, Americans had to establish themselves as worthy of such status. Worthiness, in the eyes of the founders, demanded a convincing demonstration of a ethical perspective distinct from both the British and the colonial and required proof of moral standing. Americans had to prove themselves worthy to hold to their distinct perspective and thus prove their nation to be as exceptional as their point of view on historical events. In the course of its deliberations, therefore, the Continental Congress spent a good deal of its time on matters of public reform. For example, it recommended limiting such colonial eccentricities as cockfighting, excessive theater attendance, and elaborate funeral displays. By linking behavioral reform to political independence, the Congress made an important link between the morality of private decisions and the quality of public politics. Among the "elementary books of the public right" on Jefferson's mind in his 1825 letter to Henry Lee were those of Montesquieu, who held that there is a clear and recognizable relationship between the character of a people and the quality of the nation they comprised, including the type of government they were capable of creating and sustaining. According to Ann Fairfax Withington, "the morality promulgated by Congress operated on the colonists politically" by "setting themselves up as a model. They were demonstrating to themselves and to the world that they were a moral people and that therefore their resistance to England was moral" (15). When colonists resisted the British in the name of liberty, the Americans understood their actions as informed by a moral superiority, a declarative act of nation building in which they "saw themselves as worthy of liberty, worthier than their oppressors" (17). Hence, both the moral tone of the Declaration ("the separate and equal station to which the Laws of Nature and of Nature's God entitle them") and the sense of outrage at the king's "long train of abuses" are rooted in the cultivation of a superior American morality through performative declaration. Withington concludes, "The moral movement helped bring the colonists together as a people with a will and a character" (248). The difference between people in England and people in the United States was stipulated not on ethnic or religious grounds but on categories of virtue. The Americans were ipso facto independent because they were morally superior, evidenced by their stricture, for example, that all men were created equal.

Americans' virtuous and exceptional qualities are enshrined in the Declaration of Independence's language and perspective, resulting in that document achieving iconic status in the culture at large. The declaration does a far greater thing than make a redundant claim to independence, commit an act of treason, or declare war on its mother country. It also delineates the basis of a postethnic, postnationalistic, indifferent statehood whose core definitions would be subject, eternally, to meaning established in the present condition of the nation. Statehood in the Declaration of Independence is rooted in a detachment from the kind of national loyalty located in "political bands" and transferred to the realm of perspective, in the interpretation of events, abuses, and changeable conditions. Because we see it this way, we are a separate people. Loyalty and unity are based on a sense of morally superior viewpoint and purpose. The declaration communicates vividly its sense of a people busy with a wide variety of pursuits, now grievously distracted from pressing business by a tyrant king. The declaration refers to governors trying to pass laws; legislators called to unusual and distant places only to have their legislative bodies abruptly dissolved; immigrants seeking naturalization in vain; farmers in dire need of landholdings; judges kept from administering justice; colonial offices empowered seemingly for the sole purpose of complicating private endeavors; standing armies created and billeted; exporters interfered with while trading abroad; accused persons deprived of fair trials; subordinates practicing insurrection; Indians attacking settlers; and petitioners answered with additional injury. All through the declaration's "long train of abuses," the Americans seem extraordinarily active, pursuing happiness in various ways, and getting hassled for it. They come to politics and to revolution, therefore, not for gain but because their chosen structures of attention had been tread on.

The exceptional quality of the people described in the document would seem to be based on their high tolerance for mutual existence and their ability to agree to do anything together at all. Like a pork-barrel project, the declaration's catalog of grievances probably contains items that upon closer examination would be considered trivial, inaccurate, or illegitimate, but each is included because all other interests are likely to be too distracted by their own pursuits to object to what others are doing. The various endeavors of governors, legislators, immigrant laborers, farmers, judges, civil servants, shippers, soldiers, criminals, slaves and slaveholders, Indians and settlers, are harmonized and appear on a single petition of redress by virtue of the universal, moral imperative of the pursuit of happiness. There is no master perspective here (although there is an omniscience) and no articulation of a single, guiding, all-encompassing loyalty to some overall, prior purpose, set of imperatives, or planned society. Rather, the multitude of pursuits implies a shared sense of detachment from any national or state vision; it moves away from a perspective of universality and toward one of universal, multifarious distraction. Exceptionalism is thus created in the document as an emergent self-consciousness. This is not a nation; these are the purposefully united states.

The appeal, in the first sentence of the declaration, to the "opinions of mankind" might be understood as an appeal to what were understood as universal moral standards of judgment; but more specifically, the appeal is to a universal desire to be left alone and not be tread on by predators. These standards could be found in the elementary references common to all literate people and reflected the accepted wisdom of the era. The opinions of mankind signal what Allen Jayne calls "the moral sense of the people of the world." The plain and certain reference is to the sense that "ordinary people's moral sense could be relied upon for moral judgments because it was, for the most part, unobstructed and unobscured by interested or selfish designs on the colonies" (75). The appeal to moral universals represented an attempt to depoliticize this radically political performance by shifting the attention of the listener or reader, or by framing the political action it called for in language that no "decent opinion" could deny. American colonists reading or listening to the document could thus be distracted from the awful burden of political revolution by focusing their attention on the moral imperative of their cause and their own desire to have no part in such affairs of state.

As a result, the Declaration of Independence inaugurates a U.S. public philosophy that has remained remarkably constant over two centuries. Tying political processes to moral improvement, to encourage people to pursue private happiness as a way to make society better, results in a highly personalized sense of politics and a highly revisionist style of social engagement. Americans tend to make political issues of personal behavior (theatergoing and cockfighting in the eighteenth century, smoking and sexual preferences in the twentieth), and they are equally prone to engaging in moral crusades to improve the behavior of their fellow citizens through public pressure. What are strictly personal—that is to say, nonpolitical—issues in other countries are profoundly political in the United

States. Furthermore, the declaration's conclusion about the necessity of ending political bands also inaugurates a consistently revisionist culture in which meanings are seldom if ever settled once and for all and where certainty is a very rare commodity. The single idea that the U.S. public philosophy cannot incorporate is the idea that one idea—one program, one policy—is beyond compromise or closed to amendment. The busyness of the people depicted in the declaration implies a kind of openness and tentativeness, a sense that there is a great deal of unfinished (political, capital, personal, and textual) work in America moving these distracted people around. Laws must be passed and enacted; trade must continue; settlers, servants, exporters, and criminals all have their pursuits—and the work will never be done, and if it is, it will need to be revised. All this takes time. National confidence in the workings of these processes hinges on a faith in time's abundance—time enough for personal matters and for the contingencies of public duties. American public philosophy, the implied perspective of the declaration, depends on this range of unfinished personal and political business; from the exporter's dockside office to the Continental Congress, everyone is awfully busy, fully immersed in private distractions when not called to attend to public matters.

IV. INCOMPLETION; OR, WHY IT MAKES SENSE TO KEEP RENEWING ONE'S DISTRACTIONS

The preamble to the U.S. Constitution makes the public philosophy inaugurated in the declaration into national law: "We the People of the United States, in Order to form a more perfect Union, establish Justice, insure domestic Tranquillity, provide for the common defense, promote the general Welfare, and secure the Blessings of Liberty to ourselves and our Posterity, do ordain and establish this Constitution for the United States of America." The people depicted here have met the goal of establishing a constitution, a written document, for themselves so that the union of their states will be more perfect. The intended causes of the creation of the Constitution (perhaps stated out of respect for the opinions of mankind) include what one would expect—justice, law and order, national defense, social welfare, and liberty—but also include the seemingly redundant object of forming a more perfect union. However, the redundancy dissipates in the context of U.S. public philosophy. If a natural law and a cause of separation exist prior to the Constitution, they may be traced to the declaration's very important "self-evident" truth that in this society, the people pursue happiness. The ultimate experience of happiness would logically be located in some sort of perfection. As such, this Constitution does not so much establish a nation as it establishes procedures for the nation, or the union, to continue its attempt to become "more perfect" through the pursuit of what brings its constituent members pleasure. Terence Martin has pointed out that neither the declaration nor the Constitution even provides a name for the nation they purport to establish, as both are distracted by other, more pressing priorities. "The burden of the Declaration is to convert a group of united colonies into a group of united states—just as that of the Constitution is 'to form a more perfect Union' among 'the United States of America'" (29). In this sense, the people have not formed a nation but written a manual, a textbook, and only through the application of that text do they become the "United States of America," which is, in essence, a kind of working title for a nation in progress. The states already exist as entities; the manual of procedures moves the states into a single union or nation.

When "we" the people of the present read the Constitution and consider the laws established by "we" the people of the past, how are we to align our own wills and desires for "more perfect" conditions with what has been established? The U.S. public philosophy contains an implicit interpretive imperative. The nation is rooted in texts, and texts require interpretation for ideological consumption and political praxis. The editors of a collection of American documents intended for classroom use find that so much in the United States "depends on how fundamental concepts—'freedom of speech' for example, or 'cruel and unusual punishment'—are interpreted, [and] their meaning is not eternally fixed" (Fossom and Roth 13). The meaning of these central phrases has changed dramatically over two centuries. Some are still under contention. *Equality* has meant (a) the equality of white, male landholders; (b) all white males; (c) all males; and (d) all males and females. Today, the civic and workplace implications of all men and women being "created equal" is still a matter of great contention, with no clear solution immediately available for consensus. The significance of the phrase is agreed to, while its meaning is deferred and debated in corporations, government agencies, and courts. "If the outcome precludes stagnation, it also entrenches a paradox: Americans are kept at odds by the very things on which they agree" (Fossum and Roth 13). The importance of reading and interpretation methods, not simply as private endeavors but as matters of public participation, are thus absolutely vital to existence as a fully functioning U.S. citizenry. And these practices require time in the present for contemplation, to sort consequential from inconsequential matters. A certain portion of consciousness—and that portion is subject to a wide variety of competing interests—is properly reserved for public affairs. However, attention is not easily controlled when so many media make plays to distraction and substitute the illusion of gathering information for the experience of assessing significance. Interpretation in America is a cultural act, from the political realm of judicial decision making to the individual's reading of "pursuit of happiness" against the community's desire for a more perfect union. Wholly abdicating the act of interpretation or leaving it to others surrenders active sovereignty, making happiness indistinguishable from selfishness. And yet paying too much attention could prove even more destructive.

Throughout the country's history, Americans have disputed the meanings of the ideas on which the cohesion of their nation depends. The health of a democracy is measured both by its ability to reach legislative decisions and by the way in which it manages contention. If Americans were to fully attend to their central, ideological disputes—the meaning of equality, the limits of free speech, the place of religion—and attempt to settle such issues once and for all, the social fabric would undoubtedly unravel. Shall we have a national vote on affirmative action, abortion, or capital punishment? This democracy's health is linked inextricably to a distracted population whose primary pursuits are private not publicly democratic, whose flag of public philosophy is more accurately Gadsden's serpent, "Don't Tread on Me," than the national stars and stripes of historic and cohesive expansion. Matters of contention in the United States center on phrases and ideas open to interpretation, and as long as they remain open, they provide the groundwork for union by delineating the structure of debate. The democracy is not based on one or another specific interpretation, in other words, but on the interpretable nature of all ideas and images at hand. Americans are united not by a particular definition of equality but by interest in the debate over the meaning of the term; Americans are brought together not by free speech and religious liberty as much as by the continuous discussion of the range, limitations, and applicability of these ideas. As a result, we often see Americans squared off in debate, each side standing unequivocally for liberty, freedom, and their own contingent interpretation of happiness.

Perhaps the strongest and clearest statement of the interpretive imperative in

U.S. public philosophy is articulated in Ralph Waldo Emerson's "The American Scholar": "Each age, it is found, must write its own books; or rather, each generation for the next succeeding. The books of an older period will not fit this" (53). Emerson links this eternal necessity for revision not to generational antipathy or change for its own sake but to the same philosophical principle that underlies the idea of a pursuit of happiness and a more perfect union—the faith (and it is indeed an article of faith) that the cosmos, the world, and the nation are unfinished entities: "It is a mischievous notion that we are come late into nature; that the world was finished a long time ago" (53). It is the task of human beings to work toward completion by effecting their world through their own pursuits and by affecting the worlds—moral, political, and personal—of others: "Not he is great who can alter matter, but he who can alter my state of mind" (65). Alteration, not continuation, is Emerson's ideal act, as it must be to one whose faith holds that the world is unfinished. Lincoln, in the Gettysburg Address, invoked the same faith. "It is for the living . . . to be dedicated here to the unfinished work which they who have fought here have thus far so nobly advanced" (Boorstin 437). The dead contribute not so much to a battleground as to a continuation of a project of perfection. In the twentieth century, among the most articulate evocations of the principle of incompletion has been made by novelist William Faulkner, who said in 1953, "What's wrong with this world is, it's not finished yet. It is not completed to that point where man can put his final signature on it and say, 'It is finished. We made it, and it works'" (*Essays* 135). The incomplete nature of the world calls for more writing and declaration, more performance, more thought regarding its nature and its construction. In his more widely known Nobel Prize acceptance speech, Faulkner defined the duty of writers and poets as being "one of the props" to remind human beings of their transcendent capacities (*Essays* 120).

The principle of incompletion produces a particular kind of burden on Americans and their public existence. Among the psychological stresses of democracy is that not only is the world incomplete but also no one's task is ever complete. The condition allows for the nation's historic record of compromise. If nothing is ever considered finished, compromise is more readily accomplished and accepted. However, it also results in an inability to ever consider a thing—a policy, a law, or a program—done. At any time, new knowledge, renewed priorities, or an intrusion of some sort on private distraction may compel the living generation to rewrite, revise, or restructure what it encounters. For many, the impression left is that the

U.S. government and public sector never accomplish much of anything and that what is accomplished is so heavily weighted with stipulations, exceptions, and revisions as to be ineffective. A sense of frustration with government action may result, producing a turn away from political interest or involvement and toward more private pursuits. Again, the process is good for the democracy, which depends on a distracted population. Should private pursuits suffer by public policy, revisions in the form of taxpayer revolts, recall movements, and voter initiatives will result. Any change or alteration cast as an improvement, a move toward a more perfect state of affairs or toward completion of a previously incomplete effort, requires no additional rationale. The idea that the world is unfinished so informs Americans' sensibility that the burden is more likely on the established policy to justify its perpetuation. The public philosophy exists as a prior sanction, and the national history attests to the utility of breaking with the past.

Public philosophies are made into rituals so that the public may witness and be reminded of their manifestation over and over again. The American public experiences its philosophy of incompletion every four years, when its presidency is renewed in a national ceremony of inauguration. Article 2, section 1 of the Constitution mandates that "[b]efore [the president] enter on the Execution of his Office, he shall take the following Oath or Affirmation:—'I do solemnly swear (or affirm) that I will faithfully execute the office of President of the United States, and will, to the best of my Ability, preserve, protect and defend the Constitution of the United States.'" Nothing in U.S. law then calls on a newly sworn president to address the nation. However, beginning with the first chief executive, every president has begun his term of office by delivering an inaugural address. From the awe felt by George Washington as he contemplated the victory of the thirteen American colonies over the world's most powerful empire ("Every step, by which they have advanced to the character of an independent nation, seems to have been distinguished by some token of providential agency" [Boorstin 192]), to Bill Clinton's recognition of the important ritualistic nature of an Inauguration ("We rededicate ourselves to the very idea of America" [Clinton 75]), each American president has used this tradition of renewal to reflect on the significance of his own ascendancy to power and to declare his plans to effect the nation anew. The inaugural address thus marks the relocation of the mission of the nation within the new president's ambitions and skills.

One hundred years after Washington's inauguration in 1789, Benjamin Harrison reflected on the tradition of the inaugural event: "There is no constitutional or legal requirement that the President shall take the oath of office in the presence of the people," he reminded his audience, "but there is so manifest an appropriateness in the public induction to office of the chief executive officer of the nation that from the beginning of the Government the people, to whose service the official oath consecrates the officer, have been called to witness the solemn ceremonial" (U.S. President 153). The oath taken in the presence of the people renews a mutual covenant.

In the course of one hundred years the textual imperatives of the American national tradition—that utterance, spoken and written, forms the basis of U.S. nationality and of U.S. national history—had solidified to the extent that the new president, in the course of outlining the principles of his administration, would include reference to that textuality as among the beliefs for which he stands as chief executive. Transcending any "constitutional or legal requirement," this new president knows that he must directly address the people to complete his investiture. If he does not make his declaration, if he fails to "put it in writing," as Americans say, then, in the workings of this ideology, it has not quite occurred. The inaugural address thus evolves into a ritualistic affirmation of the textual meaning of America itself. As a kind textuality itself, the United States is eternally incomplete, waiting always for the next renewal, dependent on future installations and installments. More than anything, this ideology of provisionality gives the United States its sense of youthfulness and innocence, even in a world largely shaped by its historic presence.

The oath of office and the subsequent, public submission of the inaugural address are, as Harrison intimates, acts of "manifest appropriateness" (or as Lincoln said at Gettysburg, "It is altogether fitting and proper that we should do this"). Structurally speaking, the first act committed by the new president is an interpretive one. The first official act by every new president is to offer a formal explication on the significance of his elevation to office and to interpret the meaning of the role that his presidency shall play in national affairs. It is no coincidence that every president invokes the guiding spirit of God. Any president would prefer that some higher authority should sanction the significance that human beings assign to their actions—the laws of nature and nature's God, perhaps—but American audiences know that such laws are open to interpretation. The invocation of God may be a required trope in the inaugural address, but the audience listens more attentively to the acts of interpretation made by the new president to see what he thinks it means for him to be standing there, as the chief pursuer of the nation's happiness.

Within a public philosophy of indeterminate meaning, words are adhered to as sources of unity, while the meaning of those words remains open to revision. Where periodic calls to attention punctuate public life with the regularity of seasons, Americans freely pursue private distractions with the assurance that in so doing they enact their public duty. Distraction as a way of life, as a cognitive trait, is easily confused with selfishness, especially when, untutored and left to predators, it manifests itself in purposeless consumerism and waste. The nation was founded on the right of a preoccupied people to pursue what interests them, what brings them happiness. The well-being and perpetuation of the democracy depends on the maintenance of a distracted people whose lives do not feel constrained by those who would enforce closure or attach ultimate significance to the ambiguities that hold the nation together. A nation based on textual sources requires heroic forms of provisionality and the eternal readiness of a people prepared, at will, to declare again the causes that impel them to and from the distractions of their divided attention. The traditions of *Don't Tread on Me* make the American political and cultural environment particularly equipped to manage the imperatives of the digital age. At the same time, privately managed distraction is challenged by powerful and seductive suggestions that such management be surrendered to automated information and communication sources. The single most valuable temporal asset we possess is the present, a resource that cannot be abdicated short of the complete and utter surrender of the will.

= TEACHING TO DISTRACTION; =
or, Education for the Hell of It

There was a time when a liberal education was a luxury, intended for leisure classes whose specialized forms of labor left time for cultured pursuits and the acquisition of knowledge. In predigital conditions, when information in massive quantities was economically unnecessary and news of change traveled at the pace of human conversation, only elite segments of society possessed the tools necessary to propagate and process textual materials. Today, the consumption of information of various kinds is a central activity in the lives of nearly the entire population. Information rarely comes from human sources unmediated by some technology designed to reach not one but millions of minds. Human beings may continue to think of themselves as individuals, but the goods and services they consume are packaged for their use as members of mass cohort groups. In particular, information produced as entertainment and designed to provide the illusion of knowing generally makes a confidence game of marketplace interactions. In such a milieu, the cognitive processes associated with a liberal education—the ability to sort the trivial from the significant, the habits of interrogation and interpretation—are not simply elite luxuries but mechanisms of general survival. Ironically, at the precise moment in history when such cognitive habits of mind are needed universally, schools are turning away from liberal learning toward more practical, skills-oriented training. Instead of places where the pursuit of happiness is protected as the right of Americans to be masters of what engages and distracts, schools—primary, secondary, and university—are moving toward a kind of singularity of purpose, hitching education to job preparation like blinkered mules to a wagon.

As schools become immersed in an environment where information competes with other goods and services for the attention of consumers, educational institutions' missions are in danger of being lost in an ethos of utility and disposability. A consumer item, whether a television set or a weather forecast, is accepted only if it is useful, and once its use is exhausted, it is thrown away or forgotten. As this ethos enters educational processes, schools try increasingly to function within its logic, and they fail. A liberal education is not a consumer item because it does not possess utility and it cannot be disposed once acquired. In fact, liberal education competes with utility because it consistently interrogates claims to usefulness and significance while existing outside the realm of such logic. Rather than defined by utilitarian ends, liberal education defines itself by processes: free exploration of ideas, systematic interrogation of present assumptions, continuous reinterpretation and assessment, and the expansion of the present by its detachment from productive activities by reflection and debate. Each of these processes reestablishes expansive notions of time by acting on time horizontally. Rather than focusing and contributing to the established direction of events and information, liberal education interferes with these matters, slows them down, and expands the space they occupy in the now. In an era obsessed with the future—with planning, forecasting, and projecting purposes—the free inquiry of liberal education appears to proceed for the hell of it, with no direction. As a result, its relevance is questioned.

How do we educate in the Age of Distraction? Education, vital to the workings of a democracy, is the transfer of stewardship from one generation to the next. Through its processes, the current generation prepares its children by providing what intellectual preparation it considers necessary for one to function in the world and to assume control over what continues to be called the American democratic experiment. As such, the most valuable aspect of education is conducted and understood in the service of the lives of citizens—their whole lives, not portions of them—not simply for careers or for training. The threats that face the educationally unprepared in the United States today are gullibility, boredom, listlessness, and vulnerability to intellectual predators. Not finding a job will land one on the welfare rolls, but the inability to manage distraction may cause disorientation and the loss of agency. Preparation for life in the Age of Distraction might include training in reading the codes of mass-media indoctrination, understanding the effects on concentration and deep thought produced by multitasking, and managing the expansion and contraction of time that result from digital environments. However, instead of preparing students to comprehend these phenomena, schools are more likely to submerge students into these matters as if they were not part of a constructed environment but a natural habitat. As a result, the purpose and integrity of educational institutions are confused, and they seem less like places to learn about the world and more like institutions that pretend to be the world. In agricultural communities, effective educational systems were built around the basic needs and rhythms of planting and harvesting crops. In communities of distraction, the best educational systems are constructed around the constancy of information and the incessant demands on our attention. Our crops are managed for us, but when it comes to negotiating the now of our cognitive experiences in the digital era, we are very much alone.

Early observers who claimed that Information Age technologies would create a virtual world made the mistake of assuming that virtuality would exist like a veneer over the actual and always recoverable linear world. They thought that the unmediated world would persist, with virtuality laid over it as one of its aspects, as a diversion. However, as we now know, distraction has laid claim to the major part of consciousness at the onset of the twenty-first century, leveling all experience into potential virtuality. Photography, television, video, and digital recording confuse physical memory with recorded materials, and the TV commercial that asks, "Is it real or is it Memorex?" poses a general question for the age. Each intellectual phenomenon we encounter appears as a layered distraction, taking us away from some other source of information or stimulation, something we may be missing as we attend to something else. As well, information understood as sourceless and destinationless revises the actual (in the sense of a socially "given") intransigent world of realities. Sourceless information transforms its antecedents by revealing—or recasting—those historical predecessors as virtualities. Once one era's understanding of reality is supplanted, its vestiges persist into the next as images. As a result, we become adept at the manipulation and consumption of simulacra so that what we have experienced virtually is indistinguishable in memory from what has affected us sensually or tactilely. What social theorists call postmodernism consists in large part of such cognitive reactions to the crisis in representation brought about by a universal sense of virtuality.

When most knowledge comes to us through images, knowledge itself is experienced imaginatively, and thus the line between imagination and reality (or between distraction and attention), so necessary to established educational structures, is blurred. What results is a profound and potentially transforming set of challenges for educational systems in the twenty-first century, particularly those in the humanities. The function of the imagination in sifting among distractions, in distinguishing between information consumed as entertainment and information that adds incrementally to comprehension, in unmasking the codes of mass manipulation employed by media sources—these skills are vital to intellectual survival in the Age of Distraction and are integral to humanistic education. There is nothing virtual about the "information environment" when one may ask with equal futility of data what one may ask of a tree: Who made that? However, while it may be impossible to control the emission and availability of data from electronic information technologies, it is still possible to devise methods by which one masters attention to such technologies—to devise control, that is, over what distracts us. But, as we have seen, when information is conflated with entertainment, we are likely to confuse distraction with paying attention. The situation calls for a renewed sense of purpose in humanities education. We need to get back to the basics—not in terms of what we know, but how we know.

Commercial and democratic processes have evolved in tandem in the United States, each developing means of capturing the attention of customer-constituents and each borrowing methods from the other. Education in the Age of Distraction, if it is to be effective, will center itself on the twin poles of corporate and government demands on attention. This does not mean that all students should be trained to hold office or do specific jobs; on the contrary, education is more properly concerned with the ways in which our attention is garnished, molded, and preyed on by various corporate and governmental functions. Marketplace calls to attention continually beckon consumers. Similarly, the nation's public philosophy contains procedures by which Americans are drawn away with periodic regularity from private matters toward issues of state. Implicitly, productive political distraction is understood as leading to social revitalization. Minds capable of managing distraction will return eventually to their socially committed focus of attention with renewed engagement, assuming that the distraction was not obstructed or preyed on by exploitative interests. Thus the necessity exists for some training or understanding of the way distraction works, especially in the most recent manifestations of this national style. Today, this national mode of being, this publicly held philosophy of "Don't Tread on Me," is consumed by the digital content of the Information Age, supplying a historic demand for access to but not dominance by calls to attentive engagement with affairs of the state and marketplace. Nonetheless, democracy depends on the capacity of its electorate to pay attention to public affairs with regularity and with particular critical scrutiny. A democratic citizen incapable of managing his distractions is at the mercy of media phenomena designed to fool him and to dull his critical capabilities.

Educators in the humanities face a new challenge in the twenty-first century: to explain how the information environment is shaping modes of consciousness. By teaching to distraction, teachers in literary studies, history, and in the humanities generally may provide their students what they will need to more satisfyingly and successfully navigate the cognitive demands of the various info-environments of the future. Using outmoded notions of what it means to pay attention, educators often confuse distraction with boredom or lack of motivation. Education in such cases is presented against distraction, as if in competition with more pleasurable cognitive predators, like video games and Internet technologies. Schooling this way, however, fails by neglecting to educate students about the workings of their minds in an information ecology. It would be like having a nutrition class where the teacher placed french fries and broccoli on the table and kept saying, "The broccoli is better," while the student sneaked the fries. It would make more sense to show what each ingested item did to the body, which is how nutrition classes are taught. Rather than countering distraction with medicinal education, we might better understand education in the humanities as an effort to empower the distracted mind toward controlling and profiting from its imaginative wanderings. Education ought therefore to be about distraction, to teach to it and not against it.

Under the pseudonym "Free," Abbie Hoffman published *Revolution for the Hell of It* (1968) to distinguish the revolutionary cultural ambitions of Yippies from the more overtly political motives of such groups as Students for a Democratic Society and the Black Panthers. As with many 1960s social and political movements, while we may reject much of the excess, a good deal of value can be distilled from the energies that produce cultural reevaluations. In his book, Hoffman questioned the viability of political revolution in the United States but defended the capability of people to alter the social landscape by influencing cognitive processes. The

Yippies were responsible for such socially disruptive "guerrilla activities" as levitating the Pentagon, throwing money to market traders on the stock exchange, and running a pig on their party slate for president. Such tactics, throughout the Yippies' existence, were designed to raise consciousness about historical and economic relations and to query the prospects for intellectual freedom in the modern world. Efforts at consciousness-raising in the cultural upheavals of the 1960s and 1970s were initial reactions to the dawn of the Age of Distraction. Rooted in the pursuit of happiness, Yippie politics stressed individuated pleasure and resistance to mass-defined diversions and political programs. Attempts to awaken minds, break out of "establishments," and alter modes of thought were early encounters with the sense that autonomy over what one attended to was coming under siege. To recapture one's power of distraction, rebellious minds turned to drug use, new forms of music and art, and alternative lifestyles and philosophies. While political activists worked against specific governmental policies, cultural activists worked to raise awareness of how minds were corralled for use by forces whose purposes they might have reason to question.

Hoffman stated the Yippie activists' aims plainly enough in the hyperbolic style of the era: "We are dynamiting brain cells. We are putting people through changes. The key to the puzzle lies in theater. We are theater in the streets: total and committed. We aim to involve people and use (unlike other movements locked in ideology) any weapon (prop) we can find. The aim is not to earn the respect, admiration, and love of everybody—it's to get people to do, to participate, whether positively or negatively. All is relevant, only 'the play's the thing'" (27). Hoffman's denial of a specific ideology is particularly American, as is the insistence on the absence of a comprehensive social doctrine and the romantic faith in cognitive transformation. His rhetoric has an extensive pedigree on this continent, where for centuries people have questioned the relation between their thoughts and the social, religious, and governmental forces that shape minds. As has been the case from Puritanism through Federalism, transcendentalism through Manifest Destiny, abolition through the New Deal, danger is thought to emanate not from a particular politics but from intellectual passivity. In the Age of Distraction, passivity emerges once again, more dangerous than ever, in the form of information consumption masking itself as informed involvement. Hoffman's sense throughout *Revolution for the Hell of It* is that political and ideological solutions as currently conceptualized were incapable of addressing, much less solving,

U.S. social ills in the 1960s. The only reasoned response was thus "dynamiting brain cells" to see what people would come up with once liberated from the modes of thought that had produced the current impasse.

One piece of Hoffman's program may be applied fruitfully to the humanities, to that portion of education concerned above all else with the cultural ambitions of the population, distinguished from its economic, military, and political endeavors. Education for the hell of it would not aim to earn the "respect, love, and admiration of everybody" and would favor educational programs that raise issues to fill the minds of students and the public at large with perspectives, inquiries, and intellectual stimulation capable of encountering universal distraction. Such education would stand against specified outcomes and define itself as an educational philosophy whose goal is to stimulate minds and to put people through changes so that the relationship between their cognitive processes and their social conditions became comprehensible to them. Teaching to distraction would liberate classrooms and educational programs from the limitations of teaching materials that someone already knows or that students can just as easily find stored in some data bank somewhere. It would also make clear the distinction between itself and preparatory education. Students taught to value and to cultivate their distracted minds, rather than struggle against these creative impulses, would experience their education with the confidence that their minds would one day constitute the real world of work and social policy. But the phrase *education for the hell of it* declares that education in the humanities is not for anything other than the meeting of human cognitive needs and preserving traditions of free intellectual inquiry. To fulfill such a mission, educational institutions may need to declare their independence from immediate state and corporate concerns and align themselves more specifically with the intellectual demands of a distracted people.

A democracy that suffers a crisis in purpose—brought on, in part, by an inability to manage distraction—will display lines of strain in a number of areas, including a decrease in voter turnout, increased attention to scandals and sensationalist events rather than to difficult public issues, and, more to our purposes, a belief that the educational system is failing. When society becomes moribund, educational systems are faulted for not inspiring visions of a future worth working toward. At the same time, when society becomes purposeless, is may also express its self-absorption by refusing to support education, especially when critical ideas emanate from such institutions. However, a failure to support educational systems in a democracy furthers a degeneration of public spirit. Students who sense that their communities do not value them, do not value what they are doing, and have no particular need for them or their perspectives will quickly lose their interest in or commitment to educational processes. A degenerate state of education is not the cause of social decline; more accurately, it is a symptom of a society that has lost interest in the quality of the present. And given the effects of distraction, education may degenerate as it becomes more difficult to find the present in the midst of so many signals from the future. A democracy that becomes more concerned with its future financial well-being than with its present record of justice, more concerned with middle-class standards of living than with raising minimal standards for social existence everywhere, will see its educational structures reveal the strains that result from such degradations. Students lose their spirit of inquiry as parental obsessions with personal entitlement outweigh any sense of community endeavor. Placed in classrooms that feel like holding cells, students will logically rebel either through the passivity of learning nothing or the violence of teaching others about their frustrations.

When the educational system begins to fail—as manifested by increased violence in schools, lack of talented and committed personnel, apathetic student bodies, and falling test scores—lawmakers' quickest and least threatening response is to call for the measurement of outcomes and a renewed stress on preparation. Concentrating on the test alleviates the need to confront the sources of failure outside the classroom. Reverting to military models, education critics envision schools, from elementary through graduate institutions, as versions of boot camp where raw recruits are whipped into line by instructors whose charge is to drill habits and skills into all but the most recalcitrant, who may go to the actual army. The military model makes sense for military activities. But for a democracy's institutions of education, learning to follow orders and to memorize by rote could well result in a democracy devoid of creative thought processes. And worse, students berated with standardized test scores cram for expected results, learning to value ends over processes. Few modes of thought are more threatening to democratic procedures than an inflexible priority of envisioned results. Democracies progress by compromise, by the ability to capitalize on unforeseen contingencies, and such flexibility belongs at the core of humanistic educational curricula. A lesson that democracies must teach and learn again and again is that procedures are more valuable than results and that the democratic process, as slow and frustrating and self-critical as it tends to be, is the most vital possession of its people. When the schools in a democracy are in crisis, when institutions charged with producing minds capable of working within democratic structures turn their attention solely to preparation for employment and self-gratification, the crisis very likely is a symptom of a larger degradation at work in the community at large.

The Age of Distraction thrives on the future at the expense of the present. As we have seen, the now of human existence is filled in the digital era with deferred tasks in the form of messages to answer, meetings to attend, and a general sense of busyness and of things to get to. Education, while rhetorically and programmatically concerned with the future (in the sense of the future generation and the future of society overall), deals, in its day-to-day functioning, almost exclusively with the now of students' cognition. Performance in the classroom cannot be deferred; the test is today, not tomorrow; the report is due now, not next week. Education in the Age of Distraction is challenged by minds unaccustomed to anything occurring definitively in the now; all surrounding media, from television through Internet access, promise endless additional options, sequels, updates, and links. Humanities education in the Age of Distraction is charged above all else with preserving the present moment, with calling on students to do it now, to participate, whether positively or negatively, to counter assaults on the present with creative and interpretive responses. The humanities' great art and literary works share a universal function, despite their cultural particularities: to arrest time so that the human condition may be examined experientially and wholly in the present. Students educated in the humanities possess the tools to manage distraction by recognizing their sense of the present as an aesthetic project.

It is not surprising, though, that when faced with a problem of great magnitude, experts and professionals in the field will look first to systemic fixes. Maybe the job isn't being done because someone is doing the job wrong. A sense of purpose is more difficult to manufacture than a new lesson plan or a new pedagogical approach, so plans and approaches receive much attention in times of educational reform. However, administrative procedures are not incidental to educational programs, and what seem to be positive and logical changes may unwittingly alter whole institutions' purposes and priorities. Skills are the easiest part of an education to measure (Well, can she build a spreadsheet? Can he memorize a body of factual materials?), but the formal institution of outcomes measurement tends to increase the percentage of job preparation in education and further the degeneration of democratic and humanistic educational purposes. It is very difficult to measure whether a student has had the necessary combination of creative stimulation, scientific training, and exposure to intractable problems, say, to contribute to the search for a cure to major diseases, to see the folly in some corporation's strategic plan for the coming decade, or solve the dilemma of generational dependence on welfare. It is very difficult to measure such things because no one knows the combination. If we knew the ingredients to an education that would assure the cure for certain diseases, prevent corporate folly, or reverse economic inequities, we would no longer have that disease to contend with, such high incidence of corporate blunder, or generations of impoverished families. The real problems of a democracy, in other words, are untouchable by measuring educational outcomes. If we knew the solutions, we would teach them. (Then again, if we had the solutions, we would not need to teach them.) Educational experiences ought thus to be confrontations with intractable dilemmas, not the solutions to solved problems. A confused, conflicted, and distracted graduate is far superior to a secure, smug, and passive one.

The primary danger of outcomes measurement in education, then, is not simply that it may collapse the humanities into the necessities of preparatory education. More seriously, outcomes measurement may lead in practice to the further entrenchment of the erroneous belief that educators already know what they need to know and what they need to impart to the next generation. The assumption that the current adult generation possesses adequate knowledge and insight and that a secure future depends simply on the efficient transference of that knowledge to younger, uneducated others, is a profoundly flawed logic. It is also dangerously arrogant. "Too much time is wasted because of the assumption that methods already in existence will solve problems for which they were not designed," according to Thomas Nagel; "too many hypotheses and systems of thought in philosophy and elsewhere are based on the bizarre view that we, at this point in history, are in possession of the basic forms of understanding needed to comprehend absolutely anything" (10). Not many lessons are as clear in history as the one that assures us that what seems a profound limitation in one era (such as the inability to build a machine that will fly or to envision a racially integrated social realm) will become quite common in a subsequent age. Furthermore, what we know well in one era or how we understand the world today may impede our comprehension of the direction we have already begun to take. How many people drive to work these days in a *horseless* carriage? It took at more than one generation educated by its elders to realize that the invention of the automobile was far more significant than a simple improvement on horse-driven transportation.

College faculty members everywhere commonly complain that their students are not prepared for their classes or that "today's kids don't know anything"— meaning usually that they do not know what contemporary adults know or think they already knew at the same age. There has probably never been a time when teachers did not see the decline of civilization in the faces of their new students. The Socratic paradigm of interrogating flaws in the minds of the young remains strong, and professors are accustomed to and probably need the posture of facing "unprepared," "unmotivated," and "ignorant" students. The difference is that Socrates, put to death for corrupting the youth of Athens, instilled the habit of inquiry into his students and, by example, taught them to question all claims to authority in their city. Furthermore, he assumed that his students already possessed what was necessary to think accurately and effectively and needed only prodding and dialogue to elicit it from within. Socrates compelled students to examine the quality of their ideas and to test the veracity of ideas in general; today, we are more likely to test students on their capacity to relay to us what we already know or what some employer wants them to be able to do. Students naturally will be unprepared to verbalize the assumptions of a flawed world with which they are reluctant to identify.

More importantly, however, is that what students know changes with each generation. Seventy-five years ago, before the presence of a copy machine in every faculty office, it was truly necessary for human minds to memorize vast quantities of information. In English class, for example, professors would commonly recite from memory extended passages as references or teaching tools. Teachers would also ask their students to do the same, assuming quite rightly that memorization was something a well-read English student would want to accomplish. There is today no pedagogical reason to memorize literary texts. One may wish to do so for pleasure or as a parlor trick, but the classroom does not depend on such practice. And as the necessity of such rote learning diminishes, the ability of anyone to actually accomplish memorization declines as well. It is always very difficult to get any animal, including people, to submit to demands when they can see no earthly reason for obeying. Hence, as aging faculty go to their graves reciting from the depths of Dante's inner circles, young scholars download the text from a Dante web enthusiast's Internet site and let their PC recite the text for them. As a result of this scenario it is not necessary to ask "How far has Western civilization degenerated?" but rather "What should these minds know instead?" and "How should they manage their attention?" Now that we have no need for memorization, for what alternative purposes have we been liberated? The challenge in this last question is far greater than the challenge of creating a universal outcomes-measurement tool.

We know from our experience with time-saving technologies that time is never saved as a result; on the contrary, the value allotted to tasks that have come to take less time is diminished. Now that photocopies can reproduce texts endlessly, the value of rote accuracy declines, and students will resist mastering it. But in what direction does value flee? Information technology is the externalization of human thought processes, and each new technology (from writing through word processing) has liberated the mind from some task that became drudgery through repetition. Mass literacy and cheap printing costs may have killed off the epic poem, but these technologies also made possible the high intricacy of the modernist (and unmemorizable) novel. In its broad view of human endeavor, the humanities possesses the particular mission not only of preserving and transferring the history of the human imagination but also of inspiring and evaluating its next moves. In response to the explosion of writing that followed the invention of the printing press, standards and procedures by which to evaluate the written word evolved in literate societies. We may debate without end which standards to employ, but a consensus has existed generally that some means of valuation is necessary so that we may at least categorize what has been written down. Now, with the explosion of unsolicited calls to attention, evaluative habits of mind are needed more than ever to avoid the sense that immense amounts of time are wasted encountering contentless, inconsequential bits of data and information.

It is always much easier to assure the continuation of what we already know than to allocate space in our intellects for what we have yet to consider and to create. The rate of change in intellectual capacities has been so great in the past five decades that many people quite logically retrench into known habits of mind, insisting on stability through such strategies as the enforcement of cultural literacy and other quantifiable bodies of knowledge. However, as information-storage capacities proliferate, as whole texts become accessible electronically and digitally, as search engines can locate characters, plots, themes, and images, cultural literacy may soon be wholly externalized and available like dictionary definitions. Under such inevitable conditions, minds are liberated from past toils—but again we ask, where does value flee in this situation? The liberal arts were never consumed by libraries and will not be consumed by the Internet; on the contrary, "humanities" will follow the mind as it explores territories revealed to it by the Age of Distraction. The quality of human life depends not so much on the perpetuation of past modes of thinking as on the imagination's ability to meet the world's residual challenges as imagined and created by its forbears. The humanities, which amount to the way in which human beings assess value, are charged with providing and maintaining the arena of these imaginative encounters. Far from marginalized, the humanities possess a central function in the Age of Distraction, an age that may well see the reinterpretation of how we understand the act of being intellectually engaged.

In a democracy, education reflects our current social agenda, our sense of the health of our democratic processes, and our vision of the future. Henry Giroux refers to "the academy as a public sphere" and calls for "the recognition that educators play a crucial role in shaping the identities, values, and beliefs of students who impact directly upon society" (150). Equally important to what students have learned is how and under what conditions they have been educated, the environment of their educational experience. Students will conflate their education with every other distraction that vies for their attention—schoolwork and radio and television and Internet and the rest of their digital environment. Schools that attempt to insert themselves into that environment will fail. Instead, schools need to maintain positions of authority over the contemporary era by instructing students in the arts of navigating within it, to instruct toward the management of thought and attention not toward competing with existing diversions. Students need evaluative skills, for example, to link attention to consequences and to recognize the central role that interpretive processes play in the forging of their destinies as human beings.

Attitudes toward knowledge, learning, and social interaction gained in educational settings and experiences serve as models for expectations and behavior in the larger world of the corporation and the public democracy. In a democratic culture, processes are a matters of public debate, often taking precedence over ends, policies, and programs. The structure of the school and the university, from class-room pedagogy (Are students encouraged to question or are they being taught things they could as easily learn on their own?) to curriculum development (Are students able to petition for new courses? Is the curriculum based on student development rather than faculty expertise or preference?) to the local, campus culture (Do students have a sense of campus ownership and responsibility?), is far more important to the development of minds than the specific content of their coursework. Long after content has faded from the mind, the experience of the educational process will continue to inform the citizen's attitude toward learning and adaptation, the state, authority, and the community.

Defenders of humanities who respond to accusations of irrelevance have lost the debate before it begins. Precisely this irrelevance to contemporary pressures signals the essential vitality and importance of literature and the arts. Education for the hell of it embraces irrelevance as the passageway leading us out of whatever maze we have created by our present dilemma for ourselves and our contemporaries. Human problems, in the corporate world, in government, and in personal relations thrive on the structures that produced them. Human beings are assemblers. They build continually toward greater complexity and marvel at their endless capacity to externalize their minds and their needs. Relevance is instilled in the process, and anything that does not contribute to the current structuring endeavor will be judged unnecessary or ornamental. Among the first casualties to this process is creativity. The creative impulse that produced the initial structure is no longer necessary once the structure is in place and undergoing expansion and perpetuation. Problems will arise, however, and they cannot be solved by repeating the processes that produced them. Corporate spokesmen are always calling for creative responses, but of course they do not mean true creation, which may require the dismantling of current systems. Creativity is consistently and quite logically judged irrelevant to systems or placed outside of their essential processes because of their tendency to want to tear down what exists and replace it with something else. The management and deployment of irrelevance ought thus to be among the humanities' more central concerns. Hence, the way in which humanities education is encountered is of particular importance.

The experience of learning—good and bad experiences—becomes part of the fabric of one's character. Do we graduate students who associate learning with passivity, or do we graduate students who associate learning with listening to others and engaging them in dialogue and debate? Students who associate learning with engagement in the ideas and perspectives of others, not simply with the authority figure at the front of the class, are more likely to be able as well as willing to engage colleagues, elected officials, and corporate powers in fruitful dialogue. What "is true for students who participate in class discussions," according to Elizabeth Minnich, "is also true for citizens who challenge corporate or government policies." And such participation is not only valuable to the learning process and to the state but also a source of pleasure: "People who 'talk back to power,' or 'question authority,' or just say out loud what they think once, tend to do it again and again, and not only because they are impelled by a pleased teacher or a moral cause, but because it feels so very good" (255). We might say they do it for the hell of it, as a way of arresting their powers of distraction.

Why would any corporate or government employer want to hire someone who could not manage his or her attention, someone completely at the mercy of the digital environment? As well, one cannot accept as educated a mind that is wholly clueless as to the nature of its powers of attention, ignorant of the ways in which information sources divert and command attention or the ways in which electronic machinery, PCs in particular, distract as well as process intellectual productivity. A mind that cannot distinguish consumption and interpretation, no matter how well trained in some task, cannot negotiate the demands of the age that produced it. As more work involves processing information, minds need facility in distinguishing when such activity is productive from when it simulates productivity and is more accurately understood as diverting or entertaining. Attentiveness is often considered a classroom issue, and the student who is not paying attention is likely to be dealt with punitively. However, in the Age of Distraction attention is more properly considered the central subject matter of every classroom in the same way that the glory of God was the proper subject matter of every medieval classroom. God's presence in the medieval era was felt throughout the university. Today, distraction reigns as godlike; it beckons continually, haunts the mind as it makes its way through its existence and measures its relation to a primary and ubiquitous force.

Education informed by serious consideration of distraction is more common at the lower levels of the system, where for example the nursery school allows the child's tendency toward distraction to guide its interests and activities. As the child gets older, and the stakes are perceived as getting higher, content overrides experience, and whether the student enjoys or finds pleasure in learning becomes secondary to the necessity that some measurable learning take place. Distraction as a guide to experience is replaced by submitting one's attention to authority, but the nature of attentiveness is seldom if ever investigated. As the student gets closer to an age of productive usefulness, the pressures against irrelevance get stronger, and distraction is likely to be interpreted as a failure to perform, indicating a less worthy job candidate. A radical critique, such as that produced by bell hooks, finds little "passionate teaching or learning taking place in higher education today. Even when students are desperately yearning to be touched by knowledge, professors still fear the challenge, allow their worries about losing control to override their desires to teach" (199). Students who focus their personal hostility on schools; who assault teachers, administrators, and classmates; who destroy and deface property and engage in self-destructive activities, are telling someone (if anyone can hear them) that their educational experience is cold, an affront to their sensibilities, and leaves them feeling cornered, unvalued, and irrelevant to the formation of their own destinies. The response to school violence has been to increase the authoritarian nature of the experience, which can only aggravate the problem. Education in the United States will fail if it does not resemble, in hooks's phrase, "education as the practice of freedom" (207). If freedom of inquiry, thought, and expression are the most valuable intellectual liberties of our democracy, then these freedoms need to be maintained as the cornerstone of U.S. educational systems at every level.

Every generation requires deep thinking, the kind of thought that is not easily distracted, that recognizes the ephemeral and entertainment nature of most information in the digital era. A commitment to managing distraction is one method of countering and correcting the alarming tendency of educational programs to attenuate their conceptions of student educational needs. *We are dynamiting brain cells. We are putting people through changes.* In the humanities, education for the hell of it would centralize the kind of intellectual playfulness associated with the imaginative arts. *The key to the puzzle lies in theater.* Literary cognition encourages a perspective on the world that is easily understood when referred to by terms borrowed from literary drama. According to Wolfgang Iser, "Staging in literature makes conceivable the extraordinary plasticity of human beings, who, precisely because they do not seem to have a determinate nature, can expand into an almost unlimited range of culture-bound patternings" (297). Repeated and habitual encounters with the demands of literary cognition assure that what is real will be recognized accurately for the contingency it represents and thus not be confused with inevitability. An attitude toward established policies as contingencies, knowing them as the way we happen to do things for now as opposed to forever, is absolutely necessary to the functioning of a democratic society where procedures rather than policies dominate public issues. The mind habitually distracted by literary employments becomes accustomed to an expansive, virtually unlimited range of the now so that the contemporary tendency to limit the potential and trivialize the significance of what occupies the moment may be countermanded. "Reading implies time for reflection," according to Paul Virilio, "a slowing-down that destroys the mass's dynamic efficiency" (*Speed* 5). Efficiency is the eternal enemy of creativity and alteration, both of which tend to act against standardization. As staging, the real is seen as something to encounter, to master, to orchestrate, and to transform when necessary but not something to which one must necessarily submit or into which one must mold one's capabilities.

The most important competency in education to distraction lies in one's sense of make-believe. "The experience of fictionally facing certain situations, engaging in certain activities, having or expressing certain feelings in a dream or fantasy or game of make-believe is the means by which one achieves insight into one's situation" (Walton 272). In make-believe situations, whether of the child's fantasy game or the world of literary representation, the mind engages in purposeful distraction by which it explores alternatives, lives for a while in another place, another world, and by implication renews its sense of its own base of operations as provisional space. Make-believe is not spectatorship but participation— one cannot engage in acts of make-believe without conscious effort, just as one cannot become distracted without, at some level, the complicity of will. The aim of make-believe is to get people to do, to participate, whether positively or negatively. All is relevant, only "the play's the thing." The crucial component of make-believe is the suspension of one's allegiance to what is real. A child's game can be ruined by the observation "That does not look like a horse!" just as teaching to distraction can be stalled by the insistence "That could never happen." For this reason, the study of the arts is crucial to democracy. In Iser's words, "literature becomes a panorama of what is possible, because it is not hedged in either by the limitations or by the considerations that determine the institutionalized organizations within which human life otherwise takes its course" (297). The irrelevance and the distracting essence of the arts and literature mark the centrality of these endeavors to effective educational programs in democratic societies. Aside from the standard rationales for its study—that it will increase moral awareness, sharpen the powers of sympathy and ethical engagement, or enhance a mastery of language—literary study in the United States is indispensable because it implicitly and often explicitly reinforces the contingent nature of the nation and the self-constructed essence of the person. Labeling such study as a luxury or an ornamentation implies a kind of surrender to that which exists as inevitable and unchangeably eternal.

Nonetheless, we do not normally set out with a mission to educate people so that they will be doomed to repeat or perpetuate the problems already facing the nation and the world. Education serves democracy only when it assures the maintenance of an informed citizenry capable of thinking intelligently about unforeseen problems. Those who were educated in the elementary schools of the 1950s and 1960s are today leading the world in the development of information technology and global markets. These are the same adults who as children struggled with what was called then the new math and who read in a new subject called social studies. Neither of these academic areas were familiar to parents, who struggled along with their children, wondering and grumbling about the usefulness of working with sets, unions, and intersections or of knowing the way Chinese households are organized. These methods of inquiry were unprecedented and had consequences unforeseen by those whose minds were stimulated not only by the strangeness of the subject matter but by the gulf it represented between their education and that of their parents. In fact, I would venture that the link between new math and microchips is less direct than that between a generation that believes it will reshape the world based on what it is learning and eventual, progressive change. Schooled generations must believe that what they are learning represents the necessity of the now and of the future, not the perpetuation of the past. Such belief is an important component of education. Students whose minds are taken seriously will come to understand and respect the power of thought. Minds occupied by trivialities and obedience will wither into submission. What they already know (or what they know everyone else already knows) seldom sparks minds, and few people lie awake at night pondering problems to which solutions have been found. The mind needs more.

In her collection titled *Obscure Destinies* (1932), Willa Cather's short story, "Old

Mrs. Harris," centers in part on the fate of Mrs. Harris's granddaughter, Vickie Templeton, the eldest child of a Colorado family in decline. Vickie receives a scholarship to attend the University of Michigan at Ann Arbor and discusses the matter with her more worldly neighbor, Mr. Rosen. Mr. Rosen asks her why she wants to go to college. Vickie answers, "To learn." Rosen presses further: "But why do you want to learn? What do you want to do with it?" Vickie responds, at a loss, "I don't know. Nothing I guess." Rosen won't let go: "Then what do you want it for?" Vickie holds her ground: "I don't know. I just want it." To Vickie's surprise, this response pleases Mr. Rosen, and his pleasure relieves the tense exchange between the two of them. "Then if you want it without any purpose at all, you will not be disappointed" (158).

The point of the exchange between Mr. Rosen and Vickie Templeton strikes to the core of education for the hell of it. Liberal education in the humanities historically takes for granted the idea of open-ended education and refers to it in various ways: knowledge for its own sake, the pursuit of learning, and the habit of inquiry. The concept is an article of faith among educators in the liberal arts, although as with most beliefs, it has its moments of doubt and trial. Some may recognize Mr. Rosen's quotation from Michelet, which Cather would use on more than one occasion in her writing: "Le but n'est rien; le chemin, c'est tout" (The end is nothing, the road is all). Mr. Rosen writes down the quotation on a piece of paper and tells Vickie to take it with her to college "as an antidote, a corrective for whatever colleges might do to her" (159). One thing college might do to her is to make her believe that distraction (the surest way to spend more time on the road) is antithetical to intellectual achievement, rigor, or accomplishment. It may also seduce her into thinking that the end (the test, the finished product, and the degree) possesses intrinsic value and represents some sort of closure on her education. For such doings, one needs powerful antidotes.

As the chair of the Department of English and Humanities at Bryant College in Rhode Island, I have worked at an educational institution that exists as a kind of microcosm of the U.S. social and economic environment. Bryant is a business-education college with the motto "Expanding the World of Opportunity." We have business majors in accounting, applied actuarial mathematics, computer information systems, finance, management, and marketing. And while we offer majors in liberal arts fields (communication, economics, English, history, and international studies), the vast majority of our students take degrees in business. In 1994 the college received national accreditation for its business programs from the International Association for Management Education (still known by its previous acronym, the AACSB). Perhaps ironically to some, the most recent AACSB accreditation standards demand fewer credit hours in business and more credit hours in liberal arts—at least half of any student's program of study must be in what AACSB calls "general education." The mandate, then, is that only half of the curriculum can be specifically "for" something (as in being for business opportunity). The other half must be for or about something else. There is no irony here, however. The national accreditation board refuses to conflate the preparation one receives in an accounting class with the education one receives in study engaged in for the hell of it.

Educational programs everywhere might learn something from the AACSB, especially those now adding preparatory programs to their curricula or seeking means of measuring the relevance of their curriculum to corporate interests. Of course, the issue is not new; Cather saw the problem earlier in the century, as have many others in previous education-reform movements. In setting its accreditation standards, the AACSB has attempted to counter a tendency in business education to overemphasize that first job at the expense of the lifelong career path, or the road with the more obscure outcome or destiny. Humanities education at its best prepares students for uncertainty by rewarding the qualities that technical training means to alleviate: provisionality, indeterminacy, and the suspension of decision. The emphasis on securing a job can only subvert the educational process by implying that education is equivalent to qualification or a hurdle to jump over. Liberal arts programs are not immune to such deterioration. For example, when graduate programs in literary studies equate the decline of the job market with programmatic failure, they trivialize their missions as educational institutions. Anyone who goes to college or graduate school to get a job will inevitably be (to quote Mr. Rosen) disappointed—in the job, in the education, or both.

A person who graduates from college with a degree in accounting either is going to be an accountant or is going to be disappointed. The certainty of the correlation between the field and the career (accounting/accountant; finance/financial analyst) makes the road and its end virtually the same. Such correlation is fine for career preparation but is antithetical to a democratic education in the humanities. Nonetheless, the argument that education is a road with an irrelevant destination is getting tougher to defend. It is becoming indefensible because the arguments for job training and professional preparation are in actuality arguments against education, against intellectual inquiry for the mass of students, and against the spirit of unfettered imagination. In an era of expanded access to higher education, the arguments are also antipopulist and elitist. And so to defend the obscure destinies of education appears to be an assault on the values of efficiency and productivity—articles of faith in the business world, in the national culture, and, increasingly, in the academy. Such thinking enforces the notion that education ought to be for something, whether a business career, a law degree, or a professorship in English. I fear that such thinking is potentially disastrous for teaching anything to anybody.

AACSB requirements that increase the liberal arts component at business schools are responding to developments in the business world, to what might be called a postjob environment. As William Bridges has argued, the brief era of lifetime employment in one job is passing away rapidly. In the postjob world, workers prepare themselves for a succession of task-oriented term assignments within multiple career paths. The assumption behind the latest AACSB educational requirements is that an engagement with the liberal arts will prepare students for the obscure destinies of a jobless world (education for the hell of it), while business education will assure competency in the present environment. The concept of the job can be traced to the nineteenth century in the United States, to the Industrial Revolution and the development of factories and managerial bureaucracies. For a generation or two, it made sense for companies to hire people with the expectation that they would stay with the organization for the rest of their working lives because tasks could be expected to remain relatively constant for one's lifetime. Much resistance arose to such work organization, and laborers resisted attaching loyalties to workplaces. As the industrial era yields to the consumer-information age, places of employment will yield to employment situations, and workers will become like small businesses themselves, moving from task to task, migrating with opportunities, following their interests and their distractions.

Ultimately, the interpretation of education as job training is rooted in and embodies an American tradition of anti-intellectualism, the antagonism toward any activity that is not practical, productive, or profitable. Democracies are volatile places, and while their procedures are paramount to their vitality, populations become impatient if procedures become interminable. The same may be said for educational programs that encourage experimentation and adventurous study—for every success, any number of failures. When you factor in the growing cost of higher education and the decline in federal funding for tuition assistance, the antagonism for purely academic pursuits is fueled by real hardship. Thus, we can expect those outside the academy to insist that education be for something. Those on the inside, however, cannot afford to capitulate, at least not entirely. A good portion of education, to maintain its integrity as education, must be for itself. As Vickie Templeton said, "I just want it." Students' obscure destinies are perhaps their most important possession, the one possession that enables their education.

The Age of Distraction makes educational reform not only necessary—it is always necessary to alter the ways in which we educate—but imperative. Educational structures must be rethought in relation to the Information Age as thoroughly as they were at their inception in agricultural communities. When public education was established in the United States, it seemed wise to fold it into the harvest calendar so that it took full advantage of those times when young people were able to attend to the development of their minds. In the Age of Distraction, education must be folded into a dramatically revised environment where bushels of information, not crops, are harvested regularly by students when they are not in the classroom. Minds are formed by the rhythms of dial-ups and megabytes, not by the seasons and the duration of sunlight. In this environment, schools that succeed in providing something of value to their students will see themselves not in competition with information and entertainment outlets but as sources of those skills and thought processes necessary both to navigate and to maintain intellectual autonomy in the digital age.

Distraction is a search for pleasure, and we tend to become distracted when what we are doing or should be doing ceases to be enjoyable. At that precise moment, when the mind becomes distracted and filled with a desire for something intellectually pleasurable, one's educational resources are tested. What comes to mind at these junctures reflects the quality and success—and the current status— of one's intellectual development, one's education. Every school system at every level should address this cognitive phenomenon. Rather than driving distraction underground, punishing students for exhibiting failures of attention, schools would make themselves more valuable by harnessing such moments for inquiry into the way students are thinking. Only after discovering the cognitive predicament in which students find themselves may teachers begin to serve their long-term interests. A mind cannot be satisfied by repetition; minds are truly satisfied only when instilled with a sense of creation, whether in the form of new ideas or responses or in anticipation of reward. As cognitive beings, we seek stimulation in the form of something other than what we are compelled to do—we seek agency, the capacity to act for reasons that emerge from within our own desires. Distraction is not irresponsible behavior but an expression of such intellectual longing. There is more than an implicit hostility toward the young in the current conjunction of career preparation and outcomes measurement in education, especially higher education. Instead of creating an educational experience marked by random, curiosity-driven inquiry where the student is granted freedom to explore knowledge without the burden of career advancement or corporate demands for productivity, educational institutions are allowing such burdens a legitimate place on the shoulders of their students.

Education in the Age of Distraction takes place in an environment where the consumption of inconsequential information is a major pastime and where participation in democratic systems requires cognizance of increasingly sophisticated forms of intellectual manipulation. Educational systems will meet the needs of students in this era by following broadly defined principles designed to maintain independence from the phenomena that form the subject matter of their studies and to place the intellectual well-being of students at the center of educational endeavors. To this end, we might envision a set of broadly defined principles for education in the Age of Distraction, designed particularly for the humanities, where education ought truly to be for the hell of it.

1. Students Must Be Allowed and Encouraged to Imagine More Than They Are Taught

Imagine a pretechnological life where one's knowledge of the world was limited to a village or a town and occasional posts from distant places. It did not take long to gather the news about one or two hundred townspeople and take in occasional reports from the outside world. One's conclusions about the nature of life, the political state of affairs, the economy, and human experiences generally were based on the data available. Now we multiply the sources of that data by ten thousand or a hundred thousand: News about millions of townspeople is available, hundreds of newspapers, television channels, Internet sites, all providing information about the world. But having more information available about human phenomena does not make it possible to think more deeply or systematically or to imagine alternatives to flawed methods. We know from statistical sampling, for example, that multiplying data in this way seldom alters conclusions. In fact, an overload of data may impede the capacity to envision reality in any other way.

In the Age of Distraction, information is no longer simply data or news; information is now a commodity, a consumer good, an entertainment experience purchasable as one would purchase clothing, recorded music, or gardening supplies. We don't "need" these items to survive but only to find comfort in the world we have created. The commodification of information and knowledge nonetheless has serious implications for education and students. Gregory Jay has examined the way in which higher education has lost its "monopolies on knowledge production, socialization, and the construction of ideology." Universities now compete with private corporations, think-tank foundations, and the mass media. Also, an array of activist groups make claims to the representation and production of knowledge. "'Scholarship' goes on at the Heritage Foundation; 'education' comes through MTV, advertising, the talk shows, and Hollywood cinema; 'ideology' and 'subject positions' are produced by social movement organizations through their mass mailings, cable shows, Internet and Web resources, and other public avenues" (33). In the social context Jay describes, colleges and universities ought to define their missions away from the information they provide and toward the transformational experience of education. In the Age of Distraction, schools have no monopoly on data or its use in producing knowledge and opinions about the world. Schools do possess a sanctuary from entertainment, corporate, and governmental uses of information, a sanctuary from the "real world" of established processes and problems. The world outside the academy is committed to what is, to the perpetuation of itself and its interests; the world within the academy is committed to possibility, to what could be, to what the mind is capable of envisioning. Schools represent the refuge of the imaginative. Students who are encouraged to imagine more than they are taught, who attend institutions of learning not to consume yet more information but to produce themselves, their intellectual lives, and the future of their communities, are not likely to confuse their education with watching television, consuming news, or surfing the net.

2. *Education in the Humanities*
Is the Citizen's Opportunity to Cultivate Freedom

The separation of the school from the world at large serves not only to maintain its own position of autonomous detachment but also to cultivate its students' intellectual autonomy. Cognitive freedom cannot be nurtured in situations where one's mind functions for the purpose of some specified end, product, or service. And unless such habits of mind are established early in one's development, they are not likely to be gained on the job, so to speak. That the social and corporate world works toward conformity is no secret; that the academy separates itself from that world is a principle that requires continuous renewal and vigilant protection. As corporate capitalism reaches its later stages of social consensus and as alternatives to it disappear, the tendency is to drive its specific needs into educational institutions under the guise of making education more relevant to corporate needs. Relevance, however, is simply a kinder word than the elimination of imaginative alternatives or the curtailment of intellectual freedom.

Most colleges and universities introduce the rhetoric of intellectual freedom at graduation ceremonies but do not thoroughly inform their curricula with the sense that an educator is not simply an authority in his field—which she must be, of course—but also a caretaker, holding the world in trust for the student who will supplant his teacher. Education that consists more of cultivation than inculcation will better prepare the student for what he really needs, which is the capacity to maintain intellectual freedom in a world that will threaten his mind with submission to sources of information and packaged, consumable interpretations. Thomas McLaughlin advocates the practiced detachment of higher educational systems so that the student is taught to continually theorize his relation to the state and the economy, to work through information and knowledge toward the premises on which social relations are constructed. Doing so assures that the student will be able to effectively construct the grounds of his own survival and cultivate his freedom within established structures. "Premises are not unalterable; they are culture rather than nature. When we think through premises we can see that there might be other premises, that there are other systems of behavior and interpretation that we might learn from and adapt. Theorizing in everyday culture is a survival tactic, a way of producing a livable personal and social negotiation with the rules in force" (McLaughlin 164). The challenge of McLaughlin's pedagogy lies less with students than with teachers, who must relinquish a good deal of control over materials and center classrooms around a pedagogy of critical distance from the state and the economies in which they exist.

3. Democracy Depends on an Anarchic Educational System in the Humanities

As students and faculty succeed their predecessors and as social priorities vary, educational systems must adjust, redirect, or occasionally overhaul their structures of operation from curricula to governance. Abbie Hoffman, from whose cultural anarchism I have constructed the phrase *education for the hell of it,* may be located in a long line of radical American individualists. He wrote, "Our message is always: Do what you want. Take chances. Extend your boundaries" (157)—the language may today be clichéd, but the educational implications remain vital. Much of what Hoffman appealed to in students in 1968 was the sense that education had become intellectually stultifying, prey to imperatives and agendas originating outside the classroom. To maintain its vitality, education in the humanities needs to maintain itself as society's most pressing distraction, pulling people away from information consumption and career work toward the kind of intellectual and creative desires satisfied only in a world apart from consumer, productive, and other economic pressures. In fact, because economic concerns form the common language of public discourse, they mount the strongest threat to productive humanistic anarchy.

In her book on higher education, Martha Nussbaum defines the U.S. educational system as "an unparalleled experiment, inspired by these ideals of self-command and cultivated humanity" (*Cultivating* 294). The U.S. tradition of liberal arts education has sought to provide but also to overpass the immediate concerns of career preparation. In addition, democratic principles have been invoked throughout U.S. history so that liberal education is construed as beneficial to all citizens, regardless of social or economic status in society at large. "We hope to draw citizens toward one another by complex mutual understanding and individual self-scrutiny, building a democratic culture that is truly deliberative and reflective, rather than simply the collision of unexamined preferences" (Nussbaum, *Cultivating* 294). Educational systems committed to citizenship and to cultivating the arts of democratic deliberation, creative intelligence, and, in general, smart populations (as opposed, say, to the creation of a socialist state or the maintenance of a religious order) may find it impossible to establish long-standing programs independent of demands made by students whose needs, knowledge, and capabilities vary from one era to the next. By definition, anarchy makes the impossible (but no less worthy) call for absolute freedom of the individual and allegiance to no established systems, a value antithetical to institutional survival. An educational program that is not committed to achieving what seems by current standards impossible is not in the business of education. Within this contradiction, where student freedom must be cultivated within educational systems, educational institutions must create—always, temporarily—their missions. There is no other way to build a democratic culture that is deliberative, reflective, and smart, marked by a creative and intellectually vital people.

4. The Chief Aim of an Educational Program in the Humanities Is Transcendence

Those heavily invested in the world of fact and reality will find the commitment to personal enrichment and the cultivation of intellectual distraction—in the humanities, in the arts, and in literature—too great an indulgence when the economic world of work demands such highly developed vocational skills. The gap between liberal arts colleges and vocational-education institutions, such as those housing business and polytechnic programs, will lessen as all institutions of higher learning negotiate between the intellectual needs of students in an age of media overload and the demands of parents who want their children to know how to do something and be employed after graduation. Faculty members are caught as well between the demands of society at large that what they teach become "information" (and thus subject to the same processes of commodification that now characterize other forms of consumer information) and students' needs to transcend such narrow definitions of necessity. Michael Apple sees this tension as "a structural contradiction between the task of distributing knowledge and maximizing its production. As the institutional logic surrounding the commodification process recuperates more and more of the daily teaching and research activities at universities within its orbit, the emphasis tilts toward the latter, while at the same time attempting to limit the former to only that knowledge which is economically 'essential' or to move other, more critical, forms of discourse to the margins. They, collectively, slowly become the institutionalized 'Other'" (102–3). Transcendence, spiritual and intellectual development, and other noncommodified experiences will lose the battle for resources when such battles are defined on the grounds of education that serves the interests of information manufacturers.

Education that defines itself as the transference of information is simply not part of a humanities education but an information service. Nonetheless, education masked this way is easily commodified electronically, and institutions may certainly profit by and meet public needs by supplying it as "distance learning" and other technological manifestations of what were once known as correspondence schools. Humanities education (as opposed to information about the humanities) is more difficult to commodify without degenerating into self-help and motivational services. To transcend means to go beyond the limits of the physical world or to reach past the material world toward something that eludes the grasp or comprehension of contemporary thought and established structures of knowledge. Humanities education in the Age of Distraction, with a concern for the student's whole life and not simply her image or written expression, requires forms of personal interaction and transcendence reserved only for the highest levels of human exchange. Such experiences defy commodification because they cannot be preformulated or predicted but possess the anarchic spirit of transcendent knowledge. As such, education in the humanities is not a distraction within the Age of Distraction but the single most vital resource for surviving its incessant demands for attention.

5. Education for Distraction
Is the Challenge of the Twenty-first Century

Managing distraction challenges individuals at the beginning of the twenty-first century in the way that managing to exist under the threat of nuclear warfare challenged this democracy in the decades after World War II. Students who spent a few minutes every week huddled under their desks while an air-raid signal howled outside learned, by associative experience if not by rote, that their education was tied somehow to their survival and that their survival was linked inextricably to their country's future. Entire systems of early warning and national defense were linked to their education, and their school held them in something far more important than a holding pattern prior to work. Students huddled under desks knew they were valued and knew that their survival was worth defending. Indeed, the adult population and its most sophisticated systems of defense seemed committed to assuring it. While the sirens drilled on, the educational system proceeded in safety, designed quite literally as a refuge from the hazards of political and social currents. Students did not spend the day in preparation to fight communism, they did not receive military training and were not asked too sacrifice their education to cold war imperatives. In the present era, conversely, contemporary urgencies enter the schools as education is confused increasingly with information consumption and curricula are filled with antidrug and other programs designed not to educate but to enforce behavior. Rather than sanctuaries from the compulsions of the age, schools succumb and fill their curricula with distraction.

The purpose of cold war–era education may have been rooted in a negation (anticommunist, to be exact), but the importance of shaping the way students thought about both their society and the forming of their intellects was at that time implicit in educational programs. While the prospects for general survival appear to be greater in the post–cold war era, an unintended result of peace has been to lessen the sense of young people's value. Where schools were once considered to be islands of safety (indeed, most were civil defense shelters), they now seem like dangerous places; many have added metal detectors. An educational institution ought to be an intellectual safety zone where minds gather to enrich their possibilities. Today it might be helpful to sound those air-raid signals again, warning against the contentless distractions of the Information Age and encouraging some urgent, meditative thought about our need to control the demands made on our mortal spans of attention.

THE PUBLIC VALUE
OF DISTRACTION

Pleasures lie on the other side of attentiveness. The physical act of reading, the focus, the posture, and the suspension may at times approach levels of pain, although this discomfort, usually no more than the awkwardness of elbows in the dark, is fleeting. What a text solicits is not attention, but distraction. The text provides leave, it asks that you walk alone, and accuse the world that possesses you, including your own cognitive resources, of insufficiency. After all, with the page in hand, you are looking for something, listening for it. If the reading fails, if distraction is not rewarded with even the slightest increase in the capacity to endure the world that commands attention, you will retaliate, and someone will suffer for it.

I

When it comes down to essences, distraction is about imagination. The imagination marks the relationship between physical existence and vision, between the weight of what is and the capacity to see beyond that essential limitation. Throughout their history on earth, human beings have recorded their struggle with physical and imaginative boundaries; we call the record of that struggle literature. Sustaining a literary or imaginative intellectual existence in the contemporary world requires vigilance. We live in an era of fact and documentation. Private space, where there are no shared files, camera monitors, media outlets, or electronic communication devices, must be worked at diligently to be maintained. Never before in human history have so many minds been focused on similar external stimuli, which makes possible mass society but which also constrains independent imaginative thought. Facts and documented reality are good things for the social order. Externalized, they provide touchstones for regularized thought and orderly behavior. Driving to the office or across town would be impossible without such phenomena. Nonetheless, the world of imaginative recourse, the world of privately maintained, nurtured, and controlled distraction, shrinks within the onslaught of public and social representational facticity. The tendency toward documented reality also infects literary production to the extent that it seems perfectly reasonable to ask how a work of imaginative literature relates to the author's own physical experience. The relationship between the imagination and reality is a troublesome one, and for many, the troublesomeness results in a hostility for the imaginary.

Most unsafe are children, whose imaginative capacities must be broken like wild horses to comply with social strictures. Parents are quick to instill the distinction in the minds of their children between what is real and what is made-up. Churches denounce the child's playful disregard for dogmatic truths, perhaps forgetting the tradition of playing the fool in Christ. As if to address their parents' insecurities, these children are taught to mouth platitudes in advance of comprehension. Children in day-care facilities must regiment their minds to conform to the operations of systemic processes while being socialized and taught how to play nice. At school, children bring their hostility against those who maim their imaginative capacities, lashing out violently against institutions more concerned with socialization than education. At home, the child says there is a monster in the basement, and the father takes her downstairs to show her there is no monster, only the oil burner, the water heater, and storage boxes. The father wishes to allay fears, which is a good thing to do. The danger in the enterprise is that if the child's monster is killed, so is her ability to project and respond to imaginary realms. On an important level the girl is correct: there are monsters in the basement—at least, there is little of comfort among the burners and boxes and much to keep hidden away. The monster of the imaginary rears its head throughout childhood, and those who are lucky enough not to have it stalked and killed by parents, schools, churches, or peers may enter adulthood equipped to comprehend what lies beyond the surface of their lives.

Any modern social organization craves facts. Mass society requires order above all else, and order depends on predictability and accountability, values that thrive in the realm of verification. That which cannot be proved by documentation is burdensome, and one can quite easily be absolutely right about almost anything and be discredited by misuse of fact. Every few years some journalist somewhere writes a moving story about teenage drug addiction or some other shameful aspect of American life, only to be discredited because some facts were made-up or the example chosen had no correspondence in actuality. The vehemence with which the media and public have responded to such inventions indicates little tolerance for flights of fancy within narrowly guarded definitions of what is real and what is not. The relation to truth is seldom at issue. Examples of hostility toward imaginative literature, especially for the relation and relevance it may have to the way people think, are widespread. Entering the realm of the imaginary means departing from the authoritative, verifiable realities, risking a loss of credibility and subjecting oneself to ridicule by those whose minds are deeply rooted in more literal cognitive modes.

Jon Krakauer, a journalist, wrote *Into the Wild* (1996), an essay about the death of Christopher John McCandless, an intelligent, college-educated young man who desired to live as simply and independently as possible in the United States. After substantial preparation and training, McCandless entered the Alaskan wilderness near Mt. McKinley in April 1992. He was found dead of starvation a few months later. Krakauer's book is an attempt to come to terms with the death of this young man, to assign to it a significance that would save this life from meaninglessness. In the course of his analysis, Krakauer mentions McCandless's love for the prose of Jack London (as well as for Tolstoy, Pasternak, and others). Krakauer refers to McCandless's "infatuation" with London, calls it naive, and lashes out at the literary as a category of experience and knowledge. Krakauer accuses McCandless of being so "mesmerized by London's turgid portrayal of life in Alaska and the Yukon" that McCandless would read and reread London's novels and short stories. "He was so enthralled by these tales . . . that he seemed to forget that they were works of fiction, constructions of the imagination that had more to do with London's romantic sensibilities than with the actualities of life in the subarctic wilderness." Krakauer articulates with precision the hostility held by the world of actuality for that of the imagination when he speculates on what McCandless remembered or forgot, voicing a familiar suspicion held toward romantic sensibilities. To secure his point, Krakauer denigrates the physical experiences of the mind that led McCandless astray. "McCandless conveniently overlooked the fact that London himself had spent just a single winter in the North," Krakauer tells us, "and that he'd died by his own hand on his California estate at the age of forty, a fatuous drunk, obese and pathetic, maintaining a sedentary existence that bore scant resemblance to the ideals he espoused in print" (44).

The journalist's suspicion of the hypnotic danger of imaginative fiction borders on the hysterical. The literary is accused of exerting a mesmerizing, enthralling effect on the young man. The literary made him "forget" the relationship between what is made-up and what is real and overvalue the imaginary as a realm of experience. The particularly insidious quality of fiction, however, is that it is not accountable, according to Krakauer. London's tales were not based on experience of actualities but were constructions of his imagination. The *real* London was a drunk (a fatuous one at that), not an adventurer, and thus whatever he imagined in his texts ought to be discredited—readers, like McCandless, should know better than to believe in or place their well-being in the realm of the imaginary. And so, to Krakauer, McCandless was a fool to be influenced by London's ideas or books.

What is the proper way to think about the relation between London, the man described by his critic as drunk and sedentary, and London's imagination, the source of the books that have exerted such a profound influence on readers around the world? Is *The Call of the Wild* less engaging, inspiring (and perhaps enthralling) because it is filled with made-up stuff never experienced by the author? As a category of intellectual existence, is the imaginary necessarily tied to the experiential? To say yes is to suggest that the mind has no right (because it certainly has the capability) to imagine beyond its experience. This is not a matter for debate but for the police. Someone would need to police the imagination so that it did not exceed its experiential bases. Perhaps the model of imaginative existence would be the television drama, the fact-based TV movie that explores an actual occurrence but takes liberties with details to make the drama more engaging. In such productions, the imagination is enlisted to serve the facts. In literary works, however, the relationship between facts and imagination is reversed, and actualities are enlisted to serve the imagination. If McCandless were truly inspired, he would not have felt compelled to experience such extreme wildness physically but would have imagined death without flirting so close to its threshold. Put another way, in the realm of literary distraction is the energy that drives human endeavor: whatever it can imagine, is. But those enthralled by representation must reject this hierarchy, and mass culture increasingly does so, seeing literary distractions as interference or temporary respites rather than centrality. The most vital interests of subordinated mass existence rely on that rejection.

Mass culture cannot tolerate idiosyncratic or highly individualized creation because it depends on common points of focus that can be documented, reported, filmed, or otherwise made into facts and ultimately packaged as commodities for sale and exchange. The imagination may be used to play with these focal points, but, like shared files in a network of computers, the points of focus must be sustained and available at all times. In this atmosphere, there is no hope for London, because he employed an unexperienced, undocumented wilderness and created details of the wild to serve an imaginative projection of the human condition. But the question remains, why do minds like London's place actuality in the service of their imaginative capacities? What distracts us away from what is already there to hold us?

To agree to the priority of fact over imagination calls into question a host of authoritative or consequential works from Holy Scripture to founding documents in U.S. history. If, in the Old Testament, God is credited with authoring (or at least inspiring) accounts of human action, must we discount his authority because until Christ was born, he had no experience of being human? The idea seems absurd, if God exists. There is also something of a tautology in asking whether Christ was real or a product of God's imagination. Nonetheless, human paradigms of creativity are rooted in ideas and myths about the creation of human beings themselves. Creationists have been challenged by evolutionary science; perhaps they will next be challenged by a logic that suggests that God could not have created humanity because he had no prior experience with such creatures. Genesis is reassuring, perhaps, in its claim that we were created in God's image. But this begs the question. All imaginative works are created in the image of their author, whose imagination is reflected in what is created. Nonetheless, it dodges the question of whether "God's image" (or the author's imagination) is limited to what God or author has experienced. Had God experienced human beings prior to Creation, then the Creation would be better known as the Repetition or the Derivation. The dominant Western paradigm of creativity rests on the act of creation ex nihilo, out of nothing, from the dust. We understand human creativity the same way. Operating in God's image, human beings have the capacity to conjure out of nothing something that had not been there before and to lead them and others to say, "That has not been done or said before."

More locally, Thomas Jefferson wrote "all men are created equal" while cognizant, certainly, of his slaves' problematic relation to that statement. Should the sentiment hold less authority? Perhaps in his imagination equality was a self-evident truth, while in actuality the facts of social class, economic struggle, and physical competition negated or seriously undermined what he imagined to be true. The human imagination has always existed in advance of physical reality, envisioning alternatives, providing outlets of escape, reformulation, and invigoration to those trapped, perplexed, and worn down by the limitations of the real. If it were not for constructions of the imagination, it is doubtful that human society would exist in any form higher than the brute animal organizations out of which the species is known to have evolved. Hostility toward imaginative realities may be rooted in fear of the unknown (those monsters!); it may also be rooted in a strong identification with contemporary social reality or in any number of psychological compulsions. Nonetheless, censoring, limiting, and discrediting the imagination amounts to nothing less threatening than an assault on the future of human existence. Human beings need to imagine alternatives, to be distracted not by canned and prepackaged attention-getters (such as mass-media entertainment) but by our own transcendent meandering. The imagination is most creatively distracted by thought processes prompted by sources experienced privately, by the silence of printed matter not tied to electrical outlets or socially structured situations. To imagine in this way is to experience what is most fundamentally human about the species: the ability to think, freed from the limitation of actuality and the tyranny of what is established.

The demand that the content of an author's imaginary work conform to or be verifiable in the content of his actual, physical experience was the late twentieth century's version of a very old argument against fiction and works of fancy in general. The unverifiable source of fictional authority and the unaccountability of imaginary creations have vexed hierarchies of power for as long as human beings have imagined alternative existences. Alberto Manguel, in *A History of Reading*, chronicles the plight of readers through the present era, where those who read have been "bullied in schoolyards and locker-rooms as much as in government offices and prisons." To those whose power is based not on imaginative insight but on force or on economic or administrative position, the resistance and resources of readers is continually threatening. Manguel argues that, as a result, "an artificial dichotomy between life and reading is actively encouraged by those in power.

Demotic regimes demand that we forget, and therefore they brand books as super-
fluous luxuries; totalitarian regimes demand that we not think, and therefore they
ban and threaten and censor" (21). Attacks and limitations on experiencing works
of the imagination and the delegation of the serious study of literature to electives
in college curricula flow from the threat such activities pose to established structures of power, authority, and privilege. Vigilance is necessary by those who would
defend their habits as readers. To those for whom the world as found is but a contingency, subject to the intervention of distractions explored in the privacy of the
mind, such defense comes easily, though it would indict the world of fact as a kind
of prefabricated wasteland. That's no doubt part of the problem some see with
free-floating imagination. Nonetheless, when professional critics defend literary
study, they too often reverse the relation between fact and imagination.

In 1936, Willa Cather wrote a letter to the *Commonweal* defending the function
of writing and reading literary texts that distracted from fact and social engagement. The letter is included in the volume *Willa Cather on Writing* with a 1930s
Left-baiting title, "Escapism." Leftist critics of that era strongly opposed writing
that was escapist and insisted that literature engage social issues. Writers such as
Cather suffered at the hands of critics who deemed their work irresponsible for
not directly addressing political and social interests. Cather, to her credit,
defended the superiority of the imagination over the contingencies of contemporary reality by confronting unabashedly the accusation that to do so was to be
an escapist. The epithet *escapist* (like *distraction*) is hurled at the imagination by
those who seek to stifle human creativity. Nonetheless, to use such terms is to misconstrue the literary relationship between the imaginary and the real. The central argument in Cather's essay is that to accuse art and literature of being
escapist is to engage in tautological thinking. "What has art ever been but
escape?" Cather asks, and what possible purpose can literature otherwise serve?
"Nearly all the Escapists in the long past have managed their own budget and
their social relations so unsuccessfully that I wouldn't want them for my landlords, or my bankers, or my neighbors. They were valuable, like powerful stimulants, only when they were left out of the social and industrial routine which goes
on every day all over the world." Cather's conclusion about the world of commerce and exchange is precise: "Industrial life has to work out its own problems,"
she asserted (21). When we discuss escape and the distractions that offer such
powerful stimulants to the mind, we are entering a realm altogether distinct from

that of landlords, bankers, and neighbors. The challenge is to articulate the function of escape in the world, or the public value of distraction.

Twentieth-century academic and professional literary criticism has been engaged in a quest for relevance that invites failure and brings displeasure and aggravation to its practitioners. College and university deans and accreditation boards want to know about outcomes. Career planners want to know about options for undergraduate majors. Colleagues in the social sciences and in history, who engage with documentation and fact and intervene in social and historical processes, implicitly challenge literary scholars to account for the relevance of their field. Few have responded by saying, "Industrial life has to work out its own problems." Instead, literary critics have gone to great lengths to demonstrate the social and political relevance of literary texts—to show how they interact with the world of fact and documentation. Of course, literary texts do interact with the social and political world, interrogating its representational tendencies, calling into question and thus foregrounding the relationship between language and reality. But when the hierarchy between imagination and fact is reversed, the literary imagination suffers. The ambitious graduate student may to go to work to prove that the past ten years of criticism on *My Ántonia* is false or windy (because times have changed) and the true interpretation is at hand—his hand. The ritual makes all criticism disposable. Or the even more ambitious assistant professor may set out to establish how the intricate puzzle of gender relations in her culture is exposed in the text, which she explicates. The problem of the text is thus solved, while social forces rush by. The quest for relevance through the deployment of pretext in literary studies has resulted in a succession of theoretical approaches, arguing everything from all the world's a text (and thus the province of formalist literary scholarship) to all the world's social ills exist in some particular text, canon, or tradition. Meanwhile, the function of literary criticism within the structured performances of public thinking has eroded steadily, and the only voices that are certain to reach the public at large are those that attack the institution and the practice of literary scholarship. Boo-Boo would ask Yogi Bear in a well-known anti-smoking campaign from the 1970s, "Yogi, why do people smoke?" And Yogi didn't know. Well, Yogi, why do people read?

Cather's letter to *Commonweal* addresses the question that every generation of literary scholars must answer, Why study literature? The implicit answer in recent decades has been anything but Cather's invitation to escape. Cather opens her letter with an account of primitive Southwest Native American women, living "under the perpetual threat of drought and famine," who nonetheless invest hours of effort to paint geometrical patterns on their earthenware jars and pots even "when they had nothing to cook in them." The urge to make the world aesthetically pleasing, according to Cather, springs from "an unaccountable predilection" of human behavior (*On Writing* 19). Nonetheless, the question of the usefulness of such activities dogs us, especially in U.S. civilization, where our attention is turned so persistently to productivity. Here is where Cather takes a less than genteel turn. All true poets are "useful," she argues, "because they refresh and recharge the spirit of those who can read their language" (20). Cather retreats quickly from the second point, referring instead to the "powerful stimulants" (21) in literature and defending literature against calls for immediate social relevance. The point she does not pursue is her assertion that whatever escape literary artists provide, they do so only for those who can read their language.

The women who painted intricate patterns on pots and jars saw past the practical utility of these items to something more—not necessarily more profound but certainly less immediate. At the very least, the etched shapes reflect a power of cognition that refuses to limit itself to water carrying, that by its artistic endeavor asserts that the limits of its attention are not met merely by toting water up the hill and toting human waste back down. At the root of their efforts may well have been the expansion the present so that the now of labor was filled with something more compelling and more lasting than physical exertion. The basis for reading the language of the arts is to recognize how their attention has been focused and then to read past representation to the cognitive quality of the impulse that produced it. The inability to read their language produces charges of irrelevance and uselessness because such readings read the text but cannot read past it. Likewise, the insistence on reading the language of art as if it were limited to its representational or performative functions leads to successive, disposable interpretive gestures, resulting in higher-level charges of irrelevance and uselessness. It is as if to say, "How nice that those Pueblo Indians decorated their pots. They must have had plenty of leisure time to make such pretty shapes."

In *Moby-Dick*, Melville invites the process of reading past literal experience on a number of occasions, extracting from everyday whaling operations richly evocative content. The greatness of this novel results in large measure from its sustained argument against attenuated experience in defense of oceans of imaginative exploration. The narrative itself reads past the business of whaling to a succession of complexities embodied in the purposeful actions of sailors. In one exemplary passage, Melville describes the whale line, tied to the harpoon. I quote this section at length to draw attention to a specific example of imaginative engagement.

> [T]he profound calm which only apparently precedes and prophesies of the storm, is perhaps more awful than the storm itself; for, indeed, the calm is but the wrapper and envelope of the storm; and contains it in itself, as the seemingly harmless rifle holds the fatal powder, and the ball, and the explosion; so the graceful repose of the line, as it silently serpentines about the oarsmen before being brought into actual play—this is a thing which carries more of true error than any other aspect of this dangerous affair. But why say more? All men live enveloped in whale-lines. All are born with halters round their necks; but it is only when caught in the swift, sudden turn of death, that mortals realize the silent, subtle, ever-present perils of life. And if you be a philosopher, though seated in the whale-boat, you would not at heart feel one whit more of terror, than though seated before your evening fire with a poker, and not a harpoon, by your side. (306)

The useful whale line is explained metaphorically, as it shares the qualities of the quiet that precedes a storm or of the unheld, loaded weapon—harmless and calm until its potential function commences. These images are enough to know the whale line and to appreciate its utility and its purpose in the narrative. But to read past the whale line is to recognize the thought processes contained within those metaphors. If all men live enveloped in whale lines, living day to day might be understood as the storm's calm prologue or the uncocked weapon, so that we might offer the interpretation that grace, dignity, and even the larger structures of civilization simply give form to human illusions of permanence. Here, with Melville's language, we begin to read past his image—but not quite. We are still at the level of the image of the whale line and its applicability to human nature, still thus at the symbolic function.

The last line of the passage invites the process of reading past its interpretive possibilities. One does not need to sit in a whaleboat and hold a harpoon to attend to the terror of the image created by the whale line. If you are philosophical, if you can read their language, then "seated before your evening fire with a poker, and not a harpoon, by your side," you will gain access to the cognitive horrors tapped by the coiled whale line. Are invisible whale lines less or more constricting for their imperceptible quality, the "silent, subtle, ever-present perils of life?" The gesture of poking the fire emerges as an attenuation of the harpoon thrust (as well as a hundred other such outward gestures) and the safety of the domestic hearth a delusion equal to Ahab's ambition of control over natural forces. Linked this way to the *Pequod*'s mission, fire poking becomes a task worthy of now-enriching, pleasured distraction, like toting water in decorated pots. The material circumstances of an action obscure but do not change the precarious nature of the chain of circumstances to which it ultimately contributes or the intellectual progeny to which it is kin. To read past the whale line is to read past whaling as a whole toward the cognitive processes that led Melville's imagination to be filled so deeply with harpoons, vessels, and madmen.

In *Moby-Dick* Melville consistently documents and reads past the utilities of whaling, and the whale-line passage is exemplary rather than exceptional. Nonetheless, to read past Melville is to contemplate the turn of his attention to whaling to capture images sufficient to the imaginative direction of the text. To read past the juxtaposition of harpoon and fireplace poker, for example, might lead one to recognize subsequent correspondences between images in the text and gestures contained within the reader's own daily performances. Thus, the existence of Ahab's quest for the great white whale exists over the facts of whaling like those geometrical patterns on earthenware pots. The women need those pots, but they don't need the geometrical designs except as an escape from the needed pots. Escape means flight from some oppression toward some relief; one escapes space and time to encounter an expansion of temporal and spatial existence, to increase the now in the face of its threatened extinction. Ahab and the crew of the *Pequod* need whales, but they don't need Moby-Dick except as an expression of their need to transcend or escape the business of whaling. The limits of this crew's attention are not met, in other words, merely by killing whales. If one's sense of the now is expansive, it contains room in it for the day-to-day functions of whaling and pot toting and for great white whales and geometrical figurations. Throughout the text, Melville probes the unaccountable predilection of human beings to project the forms and structures of their thoughts on wider canvases than circumstances provide, to become creatively, and at times destructively, inattentive to the practical content of their thoughts and the tasks at hand. The continuous movement by pueblo Indian women, up and down the cliffs, from their rock-perched homes to sources of water far below, most surely influenced the geometrical patterns of their aesthetic impulses, reified on their pot burdens. The continuous movement by New England whalers, further and further out to sea, to destroy life for commerce and for illumination, worked on Melville's mind so that every operation, from signing on to the ship to harpooning the whale, became invested with a significance to those who could read their language. Pleasure arises from the discovery that through studied inattention labor may come to possess value, even significance, far beyond our capacity to fully comprehend each gesture, each toting. In every case, the literary emerges from these forms and structures of distraction, and literary critics find their vocation in this mode of cognition.

Distraction is serious business in literature. At its very best, the literary is the art of reading past, of becoming distracted from what demands attention. It is not the study of history, the study of social conditions, or the study of any particular representation of reality but the study of how one reads past every one of these phenomena. Reading past means overpassing the mimetic detail—a whaling venture, a pattern of conflict, a social issue—to a consideration of the particular representation as imaginative stimulus within the larger system of meaning created by the work. And even then, one does not settle on the system of meaning but the creation of meaning, action, or movement of the system toward expansive meaning. One reads literature to be present at the creation and then to move past what has been created toward the something more that informs, in essence, the literary. Those who cannot read the language of literary texts, whose minds are satisfied with lesser works, or who prefer the seemingly less mediated language of history or social science may not be comfortable or find use in reading past but read, rather, for the text's mimetic or informational quality. This is understandable, as we seem to lack an articulate rationale for the study of literary writing, and much of what passes as literary criticism is really history, sociology, or cultural commentary. One result is that very intelligent people misconstrue their own craving for distraction and become blindsided by representational issues. Thus, if Jack London never did with his body what he imagined with his mind, we may be tempted to judge his writing inauthentic and thereby mistake the act of reading imaginative literature for the process of gathering and consuming information.

As for London, he makes his plea for the arts of distraction clear in *The Call of the Wild.* "There is an ecstasy that marks the summit of life, and beyond which life cannot rise," he writes, describing the state of mind of his main character, Buck. "And such is the paradox of living, this ecstasy comes when one is most alive, and it comes as a complete forgetfulness that one is alive." Wholly distracted, one experiences life at its most blissfully complete. "This ecstasy, this forgetfulness of living, comes to the artist, caught up and out of himself in a sheet of flame; it comes to the soldier, war-made on a stricken field and refusing quarter; and it came to Buck, leading the pack, sounding the old wolf-cry, straining after the food that was alive and that fled swiftly before him in the moonlight" (49). Far from its representational qualities, literature's test lies in its ability to engage, to entice the mind into a cognitive mode of distraction (a "forgetfulness") from the world of facts and documentation and physical existence. The ecstasy is craved; it signals life lived at its fullest. This is the escapism to which Cather referred. The mind seems to be drawn to it, like sleep when weary, like thinking itself when alert. The writer feels this ecstasy, this intense distraction, as she creates; the soldier finds it in the heat of battle; the pack dog as he pursues his prey—and, if he can read its language, the reader experiences this forgetfulness under the mesmerizing, enthralling power of textual stimulation. The experience of time is enriched, as lifetimes occur and are envisioned in the space of what are but moments on the clock once one does not hear its ticking.

Reading precedes comprehension, interpretation, and criticism. The question of how one reads is not a simple one, and its complexity has become increasingly apparent as reading habits shift and as literary criticism loses its public relevance. A number of literary scholars have become sensitive to the exhaustion of exegesis as a means of conveying literary value. Charles Altieri thus proposes "that we shift our attention from the relations between interpretive statements and their objects to the positions that works of art make available for reflecting on ourselves as interpreting subjects" (*Canons* 291–92). To a great extent, this proposal is met by moves to expand the literary canon and thus to accommodate the class, race, and gender expansion of the academy itself. As the ranks of the college educated (and of those teaching in colleges and universities) are no longer limited to specific kinds of people, the texts found useful in "reflecting on ourselves" have changed. However, those who justify the inclusion of newly canonized texts must take care not to rest on issues of interpretation and representation and thus turn imaginative work into social history. Altieri proposes a move away from concern with the constellation of interpretive paradigms surrounding literary texts and toward greater attention to textually based extensions of the reader's own imaginative capacities—and then to constructing a language and a means of explaining this process. "The problem for contemporary theory is to show how . . . imaginative activity can at once be assessed within a common language and have some influence on the principles adapted for those assessments" (*Canons* 16). Assessing imaginative activity is central to the mission of literary studies. The challenge to literary critics today is to explain, in common language, the way in which literary texts stimulate imaginative activity.

In a similar vein, Michael Bérubé observes that "while we academic readers have been devising more and more exacting ways of reading our texts, our worlds and our critics, the reading skills and reasoning facilities of [even the college educated mass public] have become cause for national alarm" (65). At the same time, the demand for imaginative writing has not declined. The exactitude with which literary critics scrutinize a text may not replicate and may not even approximate the experience of readers outside the academy. Or it may be that the ways in which academic readers read are poorly communicated by the rituals of critical presentation and poorly represented within the discourse of college and university curricula. If literary critics have not been concerned with the processes of imaginative stimulation but have instead allowed literary texts to serve various contextual interests, it may be no surprise that a gap has arisen between the ways and means of professional reading and the reasons that common readers crave imaginative writing. Among the more insidious perversions of the effort to cultivate imaginative activity through texts is the relegation, across the country, of literature to composition classes. Every year, thousands of college freshmen are given the opportunity to disdain literary study forever because it is packaged with the requirement that they write effective sentences and develop thesis statements. More than any curricular development, freshman composition represents the national hostility for imaginative activity. Who knows but that composition has had a hand in the degeneration of U.S. reading and writing skills by unnaturally combining the intellectual stimulation of literary work with the drudgery of completing formulaic assignments, often canned and processed by committees of textbook staffers.[1] Of course, colleges and universities whose corporate missions are at odds with imaginative activity may not even tolerate the required study of literature unless it is combined with some kind of skills training, like efficient writing.

While the inclusion of literary study and imaginative activity has steadily eroded over the past generation, the literary establishment has been busy fighting within its own ranks for control over the shrinking territory it occupies. John Guillory correctly dismisses the internecine battles within the literary establishment over canons and cores as a symptom of a much larger problem, a crisis shared by conservative and radical academics alike. Two distinct forms of "cultural capital" are pitted against each other in the processes of contemporary intellectual formation, "one of which is 'traditional,' the other organic to the constitution of the professional-managerial class." The literary establishment has not performed well in defending the place of traditional literary study (in any form, from great books to postcolonialism) within the intellectual demands of corporate capitalism. "It has proven to be much easier to quarrel about the content of the curriculum than to confront the implications of a fully emergent professional-managerial class which no longer requires the cultural capital of the old bourgeoisie" (Guillory 45). But it would be senseless to conclude that the professional-managerial class no longer requires imaginative activity, even if, within its hierarchies, the imagination must struggle as it has within all hierarchies—military, ecclesiastical, and now corporate. The literary academy should thus assess itself on the extent to which it supplies the resources needed by the professional-managerial class to cultivate and stimulate the imagination, to find the kind of distraction into which the rote-weary can escape for rejuvenation. Guillory's sense of "cultural capital," however, stresses the content over the active demands of imaginative literature. In many ways, what is read in English class is secondary to how it is read and to whether by the experience of reading there, the student has any desire to continue the activity on his own. When academic readers become entrenched over issues of what to read, they sidestep and become mournfully inattentive to issues of how to read.

II

Through the fence, between the curling flower spaces, I could see them hitting.
William Faulkner

This sentence is the opening line in *The Sound and the Fury*. It makes little literal sense to talk about curling flower spaces or what lies between them. Explication can fix this problem, and maybe a drawing of a pair of curled flower stems, showing how the space bordered by the stems also curls (like one of those pictures of a lamp that turns out to be two profiles facing one another) will settle the issue once and for all. But after all that, we still have this confused sentence and the question, Why was it written this way? To demonstrate how perception can be deceiving, psychologists use the trick picture with the lamp and the twin profiles. Is Faulkner's sentence meant to deceive? Unlike the profile/lamp, the sentence does not simultaneously represent two images. There is the fence, the curly spaces between the flowers, and the *I* who could see them hitting. More explication can help. The *I* is the voice of an idiot, the character Benjamin Compson, who is a retarded man, and his perceptions are thus discordant with conventional narrative methods. Of course, we would have to warn anyone who might think that Faulkner's book is an accurate depiction of the consciousness of a retarded man. Such representation would be impossible, since only the retarded man would be in a position to verify the text, and no one as retarded as Benjamin Compson can read *The Sound and the Fury*. This, then, returns us to the sentence, which, divested of mimetic qualities and any test of credibility, remains senseless. We must read past the representational sense of the sentence toward something else.

With Faulkner, there is often the problem of making sense and the challenge to read past it. Not everyone has seen it this way. In 1964 Edmund Volpe, in *A Reader's Guide to William Faulkner*, provided a number of charts, graphs, and genealogies to make sense of Faulkner's texts. Volpe's project has continued over the years, with guides, notes, interpretations, approaches, and book-length readings produced regularly to aid the uninitiated in making their way through the text.[2] There are fifteen time shifts in Benjamin Compson's narrative, and Volpe charts them. The implication is that once the reader masters the time shifts, the meaning of the text is clear. But this is not necessarily so. Knowing all the time shifts introduces an entirely new level of meaning to the text. To make schematic logic of Benjamin's narrative is to establish a level of meaning that knows what the chapter and the narrator does not know: the schematic meaning of the text's opening narrative. Even in an age when we do not talk about authorial intention, this text clearly was meant to be confusing. If we eliminate the confusion by charting the time shifts, we are damaging the text no less than if we rewrote Dickens to read: "Times the best it was of, worst the times it was of." Why would anyone do that? The revision doesn't help clarify what happens when the text is read as written. The same is true of the curling flower spaces, where making sense of the narrative destroys its effect and its meaning. Rather than making sense, the reader must read past.

Much literary criticism proceeds in the same manner as Volpe's famous (and admittedly seductive) guidebook. The critic's method is often "this means this" and "this says this but really means this" and "this is meaningless unless you know this and once you know that then this means this," or ultimately, "this says this but really means this." It is no wonder that the general readership for literary criticism—unlike that for history or the social sciences—has nearly disappeared. Who wants to be told what a book means? And worse, if you can't know one book without reading another, why read either? "You don't know about me," announces the narrator on the first page of Mark Twain's *The Adventures of Huckleberry Finn,* "without you have read a book by the name of 'The Adventures of Tom Sawyer,' but that ain't no matter." It's no matter because there is little about Huck Finn in the other book that is necessary to the experience of reading his emergence in the book he narrates. The self-defeating quality in much of literary criticism is that criticism is too often tied to explication—efforts to make the meaning plain are inevitably doomed to disposability because meaning changes with context, with readers' sensibilities, and with shifting political, cultural, and social priorities. It is no wonder that the guidebooks keep coming, each decade making sense—again—of the tales told by idiots and artists. For literary studies to reassert its value in the present, it must get past or read past the meaning of texts.

To read past—say, to read past a common literary theme, such as social tolerance—is to experience a transformation of intellectual capacities so that the mimetic question "What is this book about?" is supplanted by the reactive or compensatory question "What is this book doing to me and my distracted sensibilities?" There is no other question wholly appropriate and wholly exclusive to the literary. Consider the hypothetical example of a novel about social tolerance. To read past tolerance is to expiate a prior understanding—not merely to recognize one's own thinking but to have one's thinking suspended, distracted—so that the mind is cognizant of the parallel or the literary equivalent to the intellectual energies that tolerance necessitates. It may well be that a novel that depicts an intolerant man would be wholly beside the point. Our attention, then, might be more productively focused on "expressions of certain modes of intelligence, thought, and feeling" in the novel, expressions unavailable elsewhere, in other discursive practices (Parker 38). David Parker thus directs the reader to "the *spirit,* the *ethos* or character of a literary work's creative thinking" (38), avoiding becoming snagged on its subject matter, its theme, or its mimetic qualities. Parker is describing a kind of inattention where one willfully attends not to what is literally depicted but to the cognitive energies and structures that produced or gave rise to the representation. This kind of thinking is not always polite because it is often mistaken for irrelevance, even irreverence, and taken as a form of hostility toward pressing business. "Reading *is* a judgment," according to Sven Birkerts. "It brands as insufficient the understandings and priorities that govern ordinary life" (85). Unless one holds that literary texts are simply decorative or that the geometrical shapes serve no cognitive purpose either in the drawing or in the repeated, daily viewing, one must confront seriously the mode of attention demanded by literary distraction. To enter into the world of the curling flower spaces, one must become inattentive to the world where such utterance would interfere with business or be disruptive and decide, for the time, that that ain't no matter.

inattention
LI7. [f. IN-3 + ATTENTION.]
Failure to pay attention or take notice; heedlessness, negligence. LI7.
Lack of courteous personal attention. LI8.

The *Oxford English Dictionary* traces *inattention* to the late seventeenth century, when it signaled a "failure to pay attention or take notice" and located the resultant judgment: "heedlessness, negligence." One hundred years later, the inattentive were judged more severely and held specifically responsible for their actions, as the term since then has indicated "lack of courteous personal attention." In short, what starts as an observable human tendency ("failure to pay attention") linked to irresponsibility ("negligence"), becomes, by the end of the eighteenth century, as well a form of misbehavior, a breach of etiquette ("lack of courteous personal attention"). In the twentieth century, inattention has evolved into a diagnosable intellectual malfunction, marking its complete metamorphosis from tendency to pathology. Attention deficit disorder (ADD) is a psychological ailment commonly associated with childhood but also diagnosable in adults. Current research on the disorder, however, finds that far from indicating a deficit, the symptoms associated with ADD may signal an excess of attention. "The irony is that behavior interpreted as 'thoughtless' probably is a result of too much awareness and the desire to cope" (Cherkes-Julkowski, Sharp, and Stolzenberg 6, 9). The desire to conform to expectations and to cope with felt inadequacies produces behavioral irregularities, such as hyperactivity. Significantly, the person with ADD is more comfortable in an environment characterized by rapid shifts in stimuli and less comfortable in environments that call for sustained attention or the practice of repetitive tasks. The person with ADD may prefer his distractions provided and organized for him, in an environment at full tilt and buzzing with technologies, rather than have the burden of choosing, ordering, and managing attentiveness. The disorder could signal an advance indication of human adaptation to current trends in the social and intellectual environment. As the volume of sensory stimuli increases, the mind must spend more and more of its energies prioritizing its attentive capacities to survive.

The issue, of course, involves what one ought to attend to. The processes of socialization and education involve directing one's attention in socially accept-able, preferably productive ways. The frequency of the parent's or elementary school teacher's admonition to "pay attention!" is familiar to anyone who has attempted to direct young children, and adolescence might as well be understood as the developing human being's attempt to manage competing claims on atten-tiveness—social, biological, parental, psychic. It is no coincidence that issues of attentiveness should fall within the provinces of medical science in the late twenti-eth century—the environment produced by technology makes the phenomenon particularly problematic, and thus "inattention" joins drinking, drug use, and smoking as medical illnesses, not rational social choices. Adults are familiar with shifting modes of attention as well. Anyone who writes on a Windows-based word-processing and personal-computing system and feels the need periodically to check E-mail knows that communication technology works at cross-purposes with the art of sustained attention to single tasks. On the contrary, Windows technology specifically functions against focused attention and actively encourages a multi-tasking desktop ecology. Windows technology was established in part to allow and to encourage workers to do a number of things simultaneously. Versions since Windows 98 integrate Internet connectivity into the multitask environment. Finally, the use of multitask and Internet technology to foster reading skills acts quite explicitly against sustained attention, as web sources are predicated on the fact that alternatives or linked texts and images are always one mouse click away.3

These are not incidental matters. Indeed, communication scholars have recog-nized for decades that forms of attention lie at the core of civilization. "Attention structure is a way of understanding social organization in terms of the *structure of the system* of communication, rather than solely in terms of the nature of the signal, its content and behavioural effects"(Chance and Larsen 2). Hence, the division of attention within social groups is fundamental to the existing power structure and to the future development of the social order. For example, the way in which one's attention is deployed may define a person's relationship to the larger group or the person's relative standing within the group. Inattention may indicate anything from lack of concern to hostility, but it may also signal a position of leisured power. How others in a group react to the inattention of its various members will signal the relative social status of inattentive persons. Similarly, the function exercised by some agencies to direct our attention, such as the power of news corporations and television broadcasts to set the agenda of public concern, far exceeds the content of any particular message conveyed. In this context, what is broadcast as news is almost never as significant as the right, seemingly inherently possessed by the news corporation, to demand our attention with its agenda. News is not broadcast solely when newsworthy events occur but rather at regularly scheduled times (usually mealtimes), training the public to pay attention at set intervals and maintaining a hierarchy of exchange. Michael Chance and Ray Larsen conclude, "Because of the central role of attention in the control and coordination of social behaviour, it is probable that the social organization of attention has been *a crucial factor in human evolution*" (2).

How does a literary text direct our attention? Birkerts identifies the asocial nature of reading, where reading is understood as an implicit rejection of the demands of the community, especially the community of networked audiences. Children who prefer to read may be admonished to go play outside because it is a nice day, and bookish adults may be pressured by the structures of their lives to interact with others or engage in some demonstrably productive activity, like going to the office or mowing the lawn. Reading, though, is like walking alone, in Emerson's terms: "Whoso goes to walk alone, accuses the whole world; he declares all to be unfit to be his companions; it is very uncivil, nay, insulting; Society will retaliate" (100). In the contemporary world, where one may read texts on screen, visit web sites linked to infinite sources of textual and visual information, send and receive electronic mail, communicate in chat rooms and news groups, watch television, or perform any number of these tasks and pleasures simultaneously, the decision to read a book is not insignificant. The book suggests an individualized, self-contained consciousness, unavailable for hyperlink or on-screen notification of communication efforts. The book, especially a work of literature, demands attention in such a way as to imply the absence of need for or interest in others and a relation to the external or actual world best characterized as controlled oblivion.

Unlike a self-improvement book, a gardening manual, or study of childhood development, one cannot claim that the literary text is read to prepare for more efficient, subsequent activity. Literary reading is more accurately understood as studied, purposeful inattention, as it argues implicitly for a conception of the meaningful that may discount shared values of efficiency, including linear conceptions of time, the privileging of immediate, present concerns, and the injunction that one always be accountable in one's activities. When one reads literature attentively, one loses time, is unaware of present concerns, and is unaccountable. It is no wonder that literary employments—reading as well as writing—are seen as an affront to those with schedules to attend to or to those who seek to maintain efficiency standards. Society will retaliate. One cannot claim to need to know about Benjy's curling flower spaces or Ishmael's whale line to do one's business or properly attend to one's obligations. However, it is more likely that one cannot make these claims because literary critics have not articulated their validity.

Reading literature mirrors the kind of social rejection necessary to the creation of imaginative fiction. When readers engage in the forms of distraction demanded by the text, they participate in a ritual of inattentiveness set into the fiction by its creator, its author. The twentieth century may never have been very hospitable to literary distraction. On the eve of the century, Sarah Orne Jewett envisioned the quest for literary space as one that involved travel to the more remote regions of geographical consciousness. In *The Country of the Pointed Firs* (1896), a woman travels to Maine to find "all that mixture of remoteness, and childish certainty of being the centre of civilization" (5) that makes reading and writing possible. Hunting seclusion, she goes to Dunnet Landing to escape the social world and find the privacy and certainty necessary to creation. However, when she arrives, she becomes enmeshed in the community through her landlady's herb business and thus must struggle to create space for inattention. "Literary employments are so vexed with uncertainties at best, and it was not until the voice of conscience sounded louder in my ears than the sea on the nearest pebble beach that I said unkind words of withdrawal to Mrs. Todd. She only became more wistfully affectionate than ever in her expressions, and looked as disappointed as I expected when I frankly told her that I could no longer enjoy the pleasure of what we called 'seein' folks'" (8). The act of committing oneself to literary employments is explicitly unkind: it shows a willed lack of consideration for others and represents a breaking with the social rewards that come from attentiveness. Reading past the predicament of the writer in Jewett's passage, past the entangling seductions of the immediate world, we find that literary employments are made possible by renunciation, withdrawal, and resolution. Furthermore, these acts of unkindness and resolve are issued to gain entry into a world of certainty (although precariously vexed by uncertainty), so naturally, it would seem, the recipients of these acts would misconstrue them and suffer the kind of disappointment experienced by the abandoned Mrs. Todd.

Reading past or becoming distracted by *The Country of the Pointed Firs* brings the reader to a description of those forms of behavior produced by literary modes of thought. One cannot plead human nature here, which seems to be social. On the contrary, literary employments are located outside the natural inclinations of human beings and within the proclivities of conscious effort. Literary employments renounce natural and social connections, from the transcendent human identification with such natural forces as the sea to the sensual pleasures of being loved, cared for, and needed by someone else. To attend to the book, one must renounce the world at hand for the "uncivil" world of the distracted mind. The impulse places the imagination at the core of consciousness, as it was in childhood, before knowledge made its inroads and adult provisionality replaced the child's sense of being securely centered. With creative energies at the core and not at the fringe of consciousness, the imagination is temporarily restored to a position of cognitive dominance while the book is read and contemplated. Freed from the reciprocal obligations of sensual, interpersonal attentions, the imagination returns to its primary function, which is to provide the self with its pleasured sense of security, certainty, and reality. When the "lover of Dunnet Landing returned to find the unchanged shores of the pointed firs," she returned like a reader to a book and discovered something constant about herself. She is thus acutely aware of the predicament of her mind within competing calls to attention. By her literary activities, she makes a conscious choice to assert mastery over her distractions by returning to and repeatedly discovering the country of pointed firs.

Still, the price of discovery is the act of unkindness that makes literary employment possible. In Jewett's novel, the narrator renounces her home in New York for Dunnet Landing and then must renounce Dunnet Landing for a one-room schoolhouse, where she sets up her writing desk. The gradually more severe renunciation signals presciently the obstacles that the twentieth century would place before literary employments, as one's "Dunnet Landing" would become increasingly more difficult to locate on the psychic landscape. The need to pay

attention—to broadcasts, motorists, and communication devices—exacts a continuously rising price on imaginative impulses and foreshortens the time in which
the imagination is given reign to rule consciousness. Locating space and time that
is unavailable to electronic communication devices, immune to vehicular transportation, and safe from social interaction may simply be too great a burden for
the mind that desires to read. Reading is more likely to become an interrupted
event, and readers who are not vigilant about protecting their minds become incapable of comprehending texts that demand sustained periods of attention. It
comes as no surprise, then, to find that contemporary novels read like screenplays,
meeting the severely limited cognitive demands of much of the public. Only pre-
literate children, it seems, are exempt from the multitasking of contemporary life
and allowed to engage their imaginations over long periods; at least, that is, until
they are placed in day care or in front of a computer screen or succumb to the
pressures of socialization. And while childhood shrinks, adolescence is length-
ened, and this protracted period of powerless attentiveness results in a range of
anxieties, social disorders, and learning disabilities. While chronic inattentiveness
is tolerated in childhood only, what it represents to the mind is doubtlessly necessary to all of us.

Reading is only one form—albeit the most demanding and most encompassing
form—of sought distraction. Lesser forms allow for easier movement among tasks
and are far less individualized. In fact, most technological forms of distraction
have social dimensions structured into them. Television, magazines, sporting
events, video games, the Internet—these are all forms of distraction where the
imagination is only partially ascendant, where attention is divided but not obliterated, so as to allow the maintenance of social obligations. The packaging of E-mail
with Internet services assures that reading on-line will not duplicate or substitute
for the experience of the printed text because one is always "connected" whatever
else happens on-line. Furthermore, the simple act of being on-line makes one vulnerable to information-gathering programs or web surveillance. Reading a book
may thus be singled out as a privileged, elite form of distraction because it alone
demands individualization. The image of the singular man or woman with the
book remains a sign of individuality, and even if surrounded by a social panorama,
a bus terminal, or the blare of park music, we suspect that the reading mind has
tuned out all competing distractions for the one into which its nose is inserted.
The same cannot be said for the television viewer or the web surfer, as these minds

are continuously interrupted by invitations elsewhere. The image of the viewer is one of vulnerability; the image of the reader is one of individuality. The viewer may be sought after and sold to, while the reader sits impregnable, in full control of cognitive stimulation, as far from or close to electronic surveillance, including the telephone, as he chooses. Nonetheless, the limited forms are very popular, as they satisfy cognitive desires to enter into distracted states of mind while avoiding the responsibility of controlling one's attention. The hidden difference, though, lies in one's relationship to the demands of commercial predators and social obligations while engaged in the pleasure of distraction. Human beings do seem to crave distraction; what varies is the ability or inclination to master that need.

In a remarkable study devoted to the mysteries of attention, James Hans explains the "psychic economy that depends on inattention." Inattention is craved as a means of removing our minds from the anxiety of living in favor of alternative states of existence, suggesting "that the burdens of being a self-aware creature are too great for us to bear for any length of time." The human desire for inattention has led to "a series of social forms that are designed to take our minds off of our anxieties," resulting in "a life that inclines toward inattention rather than full awareness" (34). The extraordinary burdens of self-awareness can be and are relieved in countless ways, some of them physically debilitating. Tremendous resources are expended on forms of attention—education, preparation, development of skills—but when it comes to inattention, most people are on their own or at the mercy of entertainment industries, drug dealers, and liquor stores. Unfortunately, educational institutions fail to systematically attend to modes of inattention because, like Cather's critics in the 1930s, they equate inattention with escapism and escapism with irresponsibility. The cliff-dwelling women needed to apply geometrical patterns to their earthenware just as, in Hans's words, we need inattention because we concede that the burdens of self-awareness are too great for us to bear for any length of time. And more than that: we concede that if we are to go on toting water up the hill, we are going to have to get past the drudgery of attending to the task.

We move with distraction past our predicament and turn inattention into art— we require inattention as a mode of survival. Turning our attention to distraction signals a refusal to abdicate meta-attentive power or a refusal to surrender the power to attend to what we attend to and when we attend to it. Distraction is more than a form of anxiety release: it is a crucial component of the solution to a psychic dilemma as old as human recording. "Why are we here?" is always, in part, answered by, "So we do not have to be here." The oldest tales preserved by human cultures concern afterlives, the lives of gods, sublives, and other forms of alternative existences. The very conception of *here* implies an elsewhere, and that elsewhere is not necessarily spatial but may be (and for ages has been) a cognitive location. The relation between the two senses of here—physical and cognitive—is the province of literature. At its best, it induces a pattern of thought wholly enmeshed in a linguistic world, unconcerned for and unheeding of an actual, physical existence. Distraction, even coarse forms like mass entertainment and alcohol, is never mindless—on the contrary, it is an escape from frittering mindlessness, a mindful compensation for the dullness that enforced attentiveness produces, achieved through pleasured inattention to its sources and causes.

The heedlessness and negligence of inattention and the departure from social modes of courteous attention represented by literary endeavors provide the grounds on which we may establish a public value of literary study that is not tied to pre-textual matters such as social criticism or ideological analysis. Formalist criticism has long argued for literature's capacity to defamiliarize the real, to make it strange, as a preparation for renewed insight and greater understanding of reality. Social criticism values works of the imagination as lenses into the complexities of race and gender, historical forces, and class structures. The two modes of analysis—formal and social—were pitted against each other through most of the twentieth century, although both share a devotion to imaginative structures of knowledge. Distraction floats past all of these categories of analysis, like the student whose gaze directs his face outside, through the window in the schoolroom, although he is not looking at anything and nothing has his attention. The teacher, not the student, labels such inattentiveness *discourtesy*. Genuine inattention does not signify the absence of interest or even the presence of disinterest; rather, what we assign as inattention more likely signals an excessive intake, like sleeping scholars in library carrels. To dismiss such actions as forms of negligence is to burn the book, as it were, as an evil influence or inappropriate stimulant. The literary tradition in all of its evocations and critical paradigms is, at base, the verbal manifestation of the human need for modes of distraction. Labeled *escapist* by those with authoritarian agendas, the need to pay attention where it yields the most pleasurable cognitive return is at the heart of the literary experience. Distraction is serious business, representing something more significant than discourtesy. Our management experts, our social scientists of every stripe, would like the world to be a courteous and heedful place where rules govern human behavior—except for an occasional, structured retreat or brainstorming session. But the world is influenced more by its record of inattention than by its commitment to any particular intellectual regime.

III

Distraction signals a change in direction, according to the OED:

> distraction *n.*
> LME. [(O)Fr., or L *distractio*(*n*-), f. as DISTRACT *v.*: see -ION.]
> Diversion of the mind, attention, etc., from a particular object or course; the fact of having one's attention or concentration disturbed by something; amusement, relaxation. LME.
>
> An instance or occasion of this; something that distracts or diverts the mind or attention; distracting sounds, events, etc. E17.
>
> The fact or condition of being physically or mentally drawn in different directions by conflicting forces or emotions. L16.

The best of what we read—and here we come upon core issues of literary canon formation—specifically addresses our attention-paying capacities. The realm of the imagination is contested space because all social reality flows from it and all social potentiality depends on it. Unless one considers the world a paradise and desires no changes, all distraction, including literary representation, will be of signal importance. However, it is not mimesis that interests me but the structure of distraction built into the text. The strong, women-centered social community in Jewett's novel does not finally gain the attention so much as the juxtaposition of that community with remoteness and renunciation, suggesting the ways in which the writer's world has been defined as a redirected, absented, male world, closer to that of the sailor than the sailor's wife. Reading past representation, we come upon the cognitive links of imaginative attention and witness the recasting of the world that renunciation brings to the fore. Like a lost vessel, drifting is a necessary prelude to the redefinition or reconfirmation of one's sense of direction. And there is no greater pleasure than the sense of floating above, transcending physical and intellectual confinement, and starting or becoming anew.

Neither is there a more reliable means to become cognitively engaged in social reality, paradoxically, than by drifting from its moorings. To depart implies the possibility of return, and returning brings with it the realization of regenerative possibilities. More so, to habitually drift from the moorings of social reality encourages a frame of mind by which all social reality, including one's own self and social identity, is understood as contingency. The acceptance of contingency is the single prerequisite to social progress, whether of a radical, liberal, or conservative nature. The literary alone possesses the cognitive structures necessary to instill in earthbound minds the sensibility that apart from security dwells potentiality and that one ought to welcome rather than fear such experiences. One is surely aware when reading fiction that the world brought to one's cognitive response is elsewhere, in the meeting of the author's and the reader's imagination. Because literature is created by human acts, creation ranks prominent among its subtler implications. As long as humans create alternatives, the world, as it were, is not always. Emily Dickinson turns the sentiment to poetry:

> This World is not Conclusion.
> A Species stands beyond—
> Invisible, as Music—
> But positive, as Sound—
> It beckons, and it baffles—
> Philosophy—don't know—
> And through a Riddle, at the last—
> Sagacity, must go—(123)

Mystery thus exists like a species apart from humanity and yet as intimately involved with human experience as death itself. To live life conclusively, without access to the kind of sustained, alternative creative processes provided by the thought patterns of literary distractions, is to live as if the invisible species did not exist or had no effect on social reality. Never fully engaged in the contingency of life itself, the unfictioned mind is thus prey to whatever passing seduction captures the flitting, blinking needs of the moment.

At one point in Willa Cather's novel *My Ántonia,* the narrator, Jim Burden, agrees to sleep where Ántonia Shimerda is house-sitting because she has become fearful of Wick Cutter, the man who owns the house. When Cutter discovers Jim has replaced Ántonia, Cutter is enraged and begins choking and beating Jim. After the beating, he runs back home. "Truly, I was a battered object," he says. "I caught a glimpse of myself in the mirror. My lip was cut and stood out like a snout. My nose looked like a big blue plum, and one eye was swollen shut and hideously discolored. Grandmother said we must have the doctor at once, but I implored her, as I never begged for anything before, not to send for him. I could stand anything, I told her, so long as nobody saw me or knew what happened to me ." As for Ántonia, Jim testifies that he "hated her almost as much as I hated Cutter. She had let me in for all this disgustingness." (189). Properly explicated, this scene is read as a classic reversal of roles. Hence, we witness the education of Jim Burden as he learns about what he calls the disgustingness of the aftermath of a rape—the quality of degenerated self-valuation that follows sexual assault. Jim is thoroughly feminized by the experience, even to the point of worrying "what the old men down at the drug-store" (190) would say about it all.

To what extent can a man take the place of a woman, have her experiences, and know the world as she knows it? These are not uncomplicated questions. *My Ántonia* is written from the perspective of Jim Burden, the male narrator and substitute rape victim. And although Jim takes Cather's place in this novel, critics have differed in their reactions to this literary device (see, for example, Donovan; Fetterly; Lambert). Can a woman write through the perspective of a male narrative voice, and if she does, is her male voice merely a female voice masquerading as male? The issue is foregrounded in a brief introduction that forms part of the novel, where yet another narrator (usually understood as Cather herself) claims to have received the manuscript of *My Ántonia* from Jim Burden. But Burden is Cather's fictional creation, an experience of her imagination. Readers encountering the novel for the first time often wonder whether the manuscript really was given to Cather—that is, did she not write this book herself? The question is a good one. The novel begs its readers to read past gender, past the facts of an authorial "herself," and to ask, "If I had that body and that set of experiences, what would the world look like and feel like to me?"

Jim's response to his experience is heavily coded as socially female, as if he had been gendered "woman" by the assault on his physical self. Cather's authorial choices in this novel, including the invention of a male narrative voice that gives her the text, a text that includes his voice—which then becomes her voice—must be read past to be comprehended fully. The distractions are powerful. If a man could take a woman's place, in the logic of the Burden/Cutter scene, if he could experience the world as she experiences it, he would act and react as she does. Furthermore, the "if" is easily removed from the previous statement when the man (or the woman) is willing to suspend his sense of himself and imaginatively enter into the experience of another. Had Burden known that he was entering into the contingencies of his female friend's life (that is, not protecting her but becoming her), he might have emerged from the assault as her defender rather than her accuser. The absence of sustained distraction at this point in his life renders him psychically incapable of sympathy for Ántonia. In other words, our experiences author us, including those experiences we have when we are inattentive to our daily lives. These speculations are enforced by Cather's own effort in the novel to read past her female self and imagine, in the narrative, the position of male subjectivity. The assault on Jim is a restoration of the original reversal, where Cather becomes the male narrator and then the male narrator becomes the female victim. From what Jim experiences at Cutter's hands, one has a sense that Cather knew the risks involved in such undertakings.

The costs incurred in not attending to who we are may be psychically severe. As Stanley Aronowitz reminds us, "loyalty to the nation-state, conventionally tied to the meaning of citizenship itself, is shifted to subculture or gender, often taken as subculture" (62). Literary study in today's classroom does and always has done the work of political culture, particularly by the process of reading for recognition. Readers wish to see themselves—their sexuality, their race, their people—reflected in what they read, and, if possible, they prefer to see themselves as they would like to be: articulate, consequential, recognized. If the course of study is American literature, and I am an American, I (or someone like me) ought to be recognizable on the syllabus. Cather's text questions this pretext method of reading. Who recognizes Jim Burden, the male or the female reader? The males will say, "That's not me because the female author cannot know my experiences, even if she calls it Jim." The female reader will say, "That's not me because this particular female author must mask herself as Jim, and I am not so masked." Nonetheless, once the literary mode of cognition abandons the ability to attend to matters outside its physical boundaries, it has little left that exceeds confession. It has little that may properly be termed imaginative or literary.

As we come to accept, with increasingly less reflection, the social equation of reading and recognition, we lose sight of the fundamental value of the literary experience, which has only partially and not always to do with recognition of one's own nonliterary existence. I do not mean that literature must decenter us, make the familiar strange, or expand our horizons, make us more sensitive to others—well, I mean all of these formalist things, of course, but also something more. Literary study is the study of how one reads past the literal or, in the larger sense of this book as a whole, of how one masters (and marshals) the arts of distraction. Literary study is not the study of maleness, the study of the female voice, or the study of any particularly decentered representation or estranged reception of reality but the study of how one reads past every one of these phenomena to the structures of cognition that produce and sustain them. In this scheme, one reads to be distracted rather than reminded of where and who one is, of what the condition is, and of the daily tethers. In *My Ántonia*, Cather escaped the confines of gender as unproblematic determination of artistic production—that is, she enveloped the voice of the male and spoke it back. Cather's escape, no matter how we assess it, brings the pleasurable prospect of our own escape from confinement.

The most difficult thing for the human mind to envision is another mode of thought, a mode of consciousness that will render its own thought processes obsolete. And yet, historically, we know that such paradigm shifts have occurred with epochal regularity. Today's common sense evolves into the next era's idiocy, when what we accept as articulate expression is reinterpreted as bobbing, moaning, and slobbering. "Books that exceed our customary uses of language can teach us not just new facts—something we did not already know—but new forms of life: *something we did not necessarily know we wanted or needed to know.*" Peter Carafiol claims that such books "do not change the world. They are changes *in* the world that prompt changes in the reader" (168). *My Ántonia* is something of an intrusion, a work of fiction whose fictionality includes not only the circumstances of its emergence but the assaults it made on the mind of its author. Reading past its narrative incidents and details brings us to a realm where experience and essence cross. The relation between Burden and Cather lies between the curling flower spaces of our imaginative capacities, within the ellipses of consciousness, in the spaces left out of the current configuration of human perception and articulation.

Why else bother? Why read about what does not exist unless a sustained, intellectual encounter with "a species [that] stands apart" fulfills some psychic or cognitive requirement? A child gradually sifts reality from among the myriad stimulations of her environment. When she is done, she is an adult. If her imagination is abused, she learns to consider with disdain and to relinquish gladly the world of make-believe that previously occupied her so intently and enveloped the actual world so completely. And the world of fact and documentation does abuse the imagination, leaving scars less socially abhorrent but no less damaging in the long term than physical abuse. The need for intellectual escape does not diminish once the ability to sustain imaginative distraction is purged from consciousness. It may simply get more expensive and more damaging while becoming removed from any possibility of integration into one's social existence. Consumer goods, chronic television viewing, substance addictions—such distractions take less time, take up less room in the mind, and support rather than call into question established modes of existence. No authoritarian regime has been known to burn televisions or to shut down the market for consumer goods. Citizens with no room in their conscious lives for mindful distraction make excellent subjects.

In *The Sound and the Fury,* Dilsey says of the diminished Compson landholdings, "We aint got the room we use to have." No room, and no time: the world that emerges from the Age of Distraction is one of increasing confinement. One must travel to far off places, further than Dunnet Landing, for a psychic landscape that has room for the sustained distraction represented by Benjamin Compson. As a result, Dilsey says, Benjamin "cant stay out in the yard, crying where all the neighbors can see him" (60). The closing in of private space (or the expansion of the public, mediated community) and the technologically insistent demands to pay attention make Benjamin's voice intrusive on others who don't want to be distracted by its eternal wail. Such sounds become increasingly difficult to hear when corporate interests so relentlessly guide minds toward tentative conclusions. The meaning of human speech changes with every shift in context, and the rapid shifts of visual media result in the continual destabilization of significance, consuming more hours of viewing to achieve either the illusion of understanding or simple exhaustion. However, the multiplex environment discourages sustained thought, so nothing of significance, only ephemera, can be communicated. What is a sympathetic plea in one set of mediated circumstances becomes a pathetic annoyance in another. Rodney King asks, "Can't we all just get along?" after his beating by Los Angeles police, and the utterance eventually becomes a trope for cluelessness on the comedy club and morning radio circuit. What Benjamin Compson says, what he has to say, is of no value to these neighbors, even if they were to be told about all time and injustice and sorrow and all voiceless misery under the sun. The neighbors just don't want to know about it and are convinced that they do not have time for their inattention to be filled that way. Benjamin is illegible to the quick glance, and the conception of reality he offers invites his family and his neighbors (and his readers) to think outside their conceptual property lines, outside of what is theirs. He becomes, in short, a public issue.

Despite the private nature of reading, literary texts are social documents, and they inspire and inform public discourse. The detail of the Compson neighbors not appreciating Benjamin's wail signals an ethical issue to the reader: Should Benjamin be hidden away? The critical guidebooks make it clear that at the very least he must be explained away so that the reader's attention is not piqued to the point of exhaustion by the narrative's formulaic uncertainty. The neighbors are well within their rights to demand a quiet neighborhood; readers of the text certainly will demand the same from their own neighbors. The ethical issue finally has nothing to do with the neighbors but concerns Benjamin and his narrative. What is to be done with the man who is a completely asocial, noncomprehending, inarticulate, compulsively needy sibling, whose words require that we read another book to make sense of them? Caddy hugs him but leaves him, Quentin gets him drunk, Dilsey feeds him, Jason has him castrated and committed. Each act, alone and in certain combinations, represents a social option, and each act mirrors a critical intervention. However, to truly make room for Benjamin, we need to resist every effort to rationalize his discourse. We must renounce every hug, drink, and incision available to us as readers. Benjamin cannot be absorbed, obliterated, or edited to suit the structure of thought we bring to bear on his textual existence.

Literary encounters provoke intellectual restructuring. Martha Nussbaum has found literary texts useful on law school reading lists because "[l]iterature focuses on the possible, inviting its readers to wonder about themselves." Nussbaum is not thinking about subject matter as much as the more formal aspects of literary texts: "In their very mode of address to their imagined reader, they convey the sense that there are links of possibility, at least on a very general level, between the characters and the reader." As a result of these affinities, the reader's speculative imagination is piqued (What would it be like to be Benjamin Compson? In what ways are we like him already?) to envision alternative modes of being. The imagination of alternatives is vital to the practice of law and to the health of a democracy in general, which is why Nussbaum brings a series of literary texts to the preprofessional legal curriculum. "The reader's emotions and imagination are very active as a result," Nussbaum concludes, "and it is the nature of this activity, and its relevance for public thinking" that merits critical scrutiny (*Poetic Justice* 5). The capacity of literary study to lead the mind toward breaking through barriers of thinking, to make more space where it seems "we ain't got the room," is the pleasure of the well-flung harpoon or the perfectly wrought earthenware jug.

The value of literary study lies precisely in the function of reading past the necessary and mechanical depictions of the real and providing room for readers to do the same. The solution to an intellectual problem is seldom located in the kind of thinking that initially produced the dilemma. Academic literature programs maintain their place in higher education, frankly, because of their irrelevance to the research and teaching objectives of preprofessional curricula. Literary studies are there for the attention of those who realize that the human mind is nurtured very much by what it does not need to know. Literature distracts; it directs our attention elsewhere as a release from the confinements of yesterday's insight, the tangles of the whale line, the body of Ántonia Shimerda. From such constraints we crave the pleasures of distraction, to be "physically or mentally drawn in different directions by conflicting forces or emotions," to enter the realm between the curling flower spaces of our expectations.

Literary studies, like the humanities generally, has no programmatic mission outside of making room in the mind. Hence, literary politics are always chaotic, unsystematic, and volatile, while literary study—the experience of an expansive now—brings pleasure to those who can read its language. It feels good to clear space, to engage those senses dulled by attentiveness, to engage in studied distraction. However, once room is made, the tenants who move in are not predictable. Reading literary texts makes the mind vulnerable to intellectual and emotional challenges as well as to potential liberation. Unkind words may and often do result from literary employments. Once we have imagined ourselves as someone else— as Jim Burden, perhaps—or once we have let someone or something else control our cognitive processes, we may find our moorings revealed as whale lines, linked to monstrosities. A simple gesture, innocent as poking the fire, is thus revealed in the reading of it as participating in something demonic, in the destruction of worlds.. "So we die before our own eyes," Jewett's narrator says, as she leaves Dunnet Landing; "so we see some chapters of our lives come to their natural end" (100). Reading literature is always a judgment on the real. To chose the novel over the newspaper, the repair manual, or the television (which, even if fantasy, is punctuated regularly with commercial calls to attention), is to say no to the actual and yes (as it were) to the whale lines that connect human life to mysteries beneath its social surfaces. So we choose to die before our own eyes, over and over again. "And through a Riddle, at the last— / Sagacity must go." But even if nature will not suffice, our most durable myths tell us that from death comes new life.

The habit of reading past can be taught, but only if literary pedagogy remains distinct from most other forms of teaching, forms that rely on the importation of knowledge and on making clear what is to be known. Literary study is often sabotaged by instructional methods that call on students to read texts as sociological or psychological cases or as formalist or linguistic puzzles. These methods read but do not read past their texts. Those entrusted with teaching the study of literature, then, would serve the interests of their students better if they claimed to teach methods of distraction. The feeling of Friday afternoon to the nine-to-five worker, the anticipation of the bell to the public-school child, the embrace of the infant after the absence of the parent, the lover's eyes across crowded public spaces—these are the pleasures of welcomed distraction, promising escape from here and access to another level of existence, where muted senses are brought back to life and the now expands to embrace all time. The purpose of literary study is to make room in the mind *now* for such pleasures of renewal, so that it can read past what it knows or is expected to know and migrate to other cognitive structures of knowledge.

Literary critics use the present tense when they discuss texts because literary enactment always exists in the present, in the now. Benjamin Compson moans and slobbers eternally and holds the jimson weed for comfort. In the now, with Benjamin, are Jim Burden and Ántonia Shimerda, the whale line, Dunnet Landing, and the curling flower spaces. In a prolonged and extensive now is where literary texts place the minds of those who can read their language, not in the name of attending to the present but with the purpose of exploring underneath, beyond, and past it. And what else is pleasure but a heightened sense of one's physical existence in the now? For centuries, drugs and alcohol have aided human beings in their quest to expand experience, and specific narcotics have influenced more than one literary movement in history. Sexual pleasures awaken bodily sensations, held in check by a transit civilization whose purposes are complicated by staying in one place for very long, long enough for sensuality to overtake the pressure of commuting and clock-watching. Pornographic industries arise in the crevices of such longings. Such distractions cannot be incorporated by demands for attention—don't drink and drive and don't read and drive—but neither can the pleasures they represent be expunged from human impulses. Literary study cannot abandon the pleasures of inattention without abdicating its essence as a tradition—not as a canon but as a mode of thought. Once we read past historical fiction, science fiction, comedy, tragedy, romanticism, and the rest, the single, universal object of literary study is the present in all its limitlessness and expansion. Anything less trivializes our lives as readers, the vocations of literary scholars, and the purpose of literary study. Above all else, we read literature to extend the present, to fill the now as fully as possible, paying no attention to the tendency of human institutions and information technologies to trivialize the now by insisting that memory or management define it essentially. In the literary now, between the curling flower spaces, time will always tell what we get when we make room for the desires of literary employments. This is what feels right, feels good.

First comes Benjamin Compson, in other words, full of sound and fury, and then comes puzzled attention, like when Benjy remembers his sister, Caddy, running toward him with her book satchel to listen to what he has to say. The strength of such literary encounters has always resided, ultimately, in the mode of thought resurrected on the page.

"Did you come to meet Caddy." she said, rubbing my hands. "What is it. What are you trying to tell Caddy." Caddy smelled like trees and like when she says we were asleep.

What are you moaning about, Luster said. You can watch them again when we get back to the branch. Here. Here's you a jimson weed. He gave me the flower. We went through the fence, into the lot.

What do we want to say, what do we moan about, when we become so distracted that we read past the page? Literary recognition is not a matter of mirrored reflection but an abrupt encounter with something or someone we thought we had lost, some way of thinking possessed by someone else, in another place, at some other time. Seeming to transpire in solitude, reading is rather an intensely intersubjective act, interlocking as it does two minds, one present, and one residual, forming a continuous community amidst the roar of *isolatos* whose quest is eternally for connection to something significant. As social as reading is, it defies mass culture because it engages minds without mediation and at a pace and schedule set entirely by the reader, who controls the presence and the intensity of cognition. Reveries induced by reading travel without regard to electronic information exchange. Mass culture cannot expand the now but can only fill it, buy it some time for purchase later on. Literary employments burst the seams of the now to sustain cognitive alternatives. In the text something survives, either coarse and poisonous, like the jimson weed, or playfully welcomed, like the jouncing book satchel, dogging us through the fence and into the realm of public thinking. When we accustom our minds to the patterns on the earthenware pots, we find ourselves reading past the function of the jars, escaping into the realm of what tugs at our private whale lines and what provides when we become heedless, negligent, and inattentive.

= SPECULATION AND SURVIVAL =
IN THE AGE OF DISTRACTION

This World is not Conclusion.
A Species stands beyond—
Invisible, as Music—
But positive, as Sound—
Emily Dickinson

SPECULATING NOW

What percentage of our lives exists in the imaginary? What motivates reflection, leading the mind away and for a while drawing on something interior, a harbor from the pressing external realities of existence? Emily Dickinson likened the phenomenon to a sense that life offered more than externalities, that some species of life paralleled outward existence and was accessible through paradox, invisible but positive, a sound heard like thought but not empirically there. Outward lives are indications, not summations, of existence. The imaginary is among the mind's most vital resources, the interior existence that we all lead and that may either buttress or subvert our social lives. It is an asset we have left largely unprotected. Perhaps the most insidious aspect of the Age of Distraction is a capitulation to the authority of external sources and the denial of individual autonomy. Avant-garde philosophical movements, such as postmodernism, contribute to these denials by speculating about the end of subjectivity, the exhaustion of the individual agent as source and generator of meaning, as the keeper of its own imaginative powers. There is no surer way to attenuate the imagination than to imagine its extinction as the species beyond.

Paradoxically, the imaginary must depend on its own resources for what we know of its existence. Whatever we understand the imaginary to be relies on our imagination's ability to project it. If we know it as a parallel species, we might it find worthy of our cultivation. If we deny its existence or abdicate it to external sources, then by its own power it will be destroyed. The imagination is inextricable from the will, so much so that only it can imagine itself into being. The imagination, we might say, is something we have imagined ourselves to possess. It has no objective existence, it cannot be removed from the mind and examined. It cannot be discussed objectively; only through the subjective employment of metaphorical language, the language of figurative, poetic discourse, are we able to project its existence within us. There is a kind of faith that underlies such endeavors, a faith that led twentieth-century artists such as Willa Cather to align art with religion, claiming the two phenomena to be equivalent. This is not to say that art is to be worshiped or that artists should be revered like gods. Rather, the equivalency is seen in the acceptance without empirical proof of the existence of a species beyond, a species that depends on that faith for its manifestation as an observable, effective phenomenon. One who cannot envision an autonomous imagination, distinct from the external world and capable of withdrawing from its imperatives, will never know what it means to possess such a resource for survival.

We had a kind of sad and eccentric old man in my extended childhood family. Uncle John was my grandmother's bachelor brother who lived with her and my grandfather. When I knew Uncle John he was about seventy years older than I and had a bicycle shop in Hartford. Before that, he had been a tailor, and before that an automobile mechanic. He had a set of books, in greasy supple leather bindings, about all kinds of machines and how to build and repair them. He also had a large encyclopedia about bicycles and another with pictures of different kinds of clothing and how they were made. Those books were his professional autobiography. He would tell me, "The only kind of book worth reading is one that tells you how to do something." He should know, as his life had consisted of reading and doing. I remember watching him in the bicycle shop. He would go back and forth between that great big book with the bicycle wheel on the cover and the bike, in pieces, on the stand in the shop. He'd read a little and fix a little, read a little more and fix some more. "This is what a man does," he would say. I asked him once what he would do without the book. "Without this book, there would be a lot of broken bicycles." Once in a while he would construct a bicycle from spare parts and give it to me. These bikes were legendary in my neighborhood because they were indestructible and looked nothing like what stores sold. I could ride it up and down dirt mounds, jump curves, and crash into shopping carts and it would not so much as bend. He could never sell these hybrids, he said, because no one wanted something that did not look like a picture on a brochure. Sure, part of me longed for a store-bought bike, but those "Uncle John Specials" were part of my world—and they were the fastest things on the block.

Uncle John harbored deep suspicions and felt that his life was controlled by conspiratorial forces. When he saw me reading a novel from the Scholastic Book Club that serviced our school, he would tell me to be careful. "That book may not let you come back," he'd say, "and it's not going to help you do anything." I had visions of being sucked into my book by something like a tornado funnel, taking me away from the tiny rooms and claustrophobic structures of our suburban tract house. Uncle John had his secret life, though, in the shop, building new kinds of bicycles and carts and anything with wheels. When he was done reading and fixing, he'd tinker and build for hours. Anything he made was indestructible but fairly ugly. "A good bike is built for use, not show," he'd say. If he or anyone else in the family knew something about business, he might have invented something marketable. Sometimes he'd give one of his contraptions away to some neighborhood kid who couldn't afford a bike. The block was hit by riots at the end of the 1960s, but his shop was spared, while adjacent stores were completely destroyed. Uncle John was a fixture on that block, just as he is fixed in my mind, finally done with the small-change repairs, spending hours creating some six-wheeled tricycle able to jump city curbs. When he died, his shop was junked and his debts paid by life insurance settlements.

So it was with Uncle John in his way as it was with my father in his: go to the shop and tinker, go to bed and read for a while, duplicating and protecting the privacy of cognitive escape with physical sanctuary. The biography of a person's imagination is marked by the succession of objects to which he has attended. Attending solely to sources that instruct action places the mind in the service of what it encounters externally. Reading is tinkering: the reader is what Ralph Ellison called a "thinker-tinker," making the associative links that move thought like a bicycle chain driving wheels forward and back. Reading in the age of distraction is a strategy of survival, an aid to the mind that, if it is to possess integrity, must work to transcend the pervasive, electronic bicycle-repair manuals that continuously direct action and desire. The repair manual insists that one's attention have relevance and possess purpose, that it proceed always at the service of some guiding plan of action. Read a little, do a little; read a little, do a little. Imaginative resources and the arts generally have no such relevance and serve no aggregate purpose in the physical world. Hence my Uncle John's ambivalence. Associating the imaginary with phenomena that controlled and directed him, he suspected all external sources of manipulation. And yet there he went, inventing bicycles to withstand the conditions of city and suburban streets and the abuse of the children who would ride them. To us, he was immensely amusing, with his ideas about the conspiracies and dangers around us and the "junk" that emanated from mass-production bicycle lines, designed to fall apart and (he said) kill us. Today it is not so amusing, as more than bicycles roll into our minds, dimming our awareness of the mind's capacity to shape the world it inhabits.

The relationship we hold to reality is called the imagination. The stronger one's sense of the imaginary, the more control one may exert over the extent to which reality affects one's own "species-beyond" existence. A powerful imagination knows realties to be contingencies: though no less real for their provisional nature, they possess no privileged status in the interior processes of speculative thought. Artists, as a result, can produce works of the imagination capable of becoming real things. Moby Dick may not have existed empirically, but that whale occupies more space in "the real" than most of its species and in that way exists. The less active one's imaginative realm, the greater the thrall to which one submits to such externalities as social pressures, intellectual predators, and the package deals of digitized escape. An imagination that serves reality demands proof, as it is wed to the empirical world for its functioning. Cultivating the imagination in the Age of Distraction requires an act of will. The world is populated by millions of Uncle Johns who can find no outlet for their desire to shape the conditions of their lives and so feel besieged by the manuals on which they depend for survival. Religious systems, aesthetic movements, thousands of years of what we call the humanities testify to the insufficiency of the external world to hold the species and provide the mind the sense of completion it desires. "This world is not conclusion," and in the other, the species thrives on the fissures of consciousness that erupt in response to the nature of inconclusive existence.

The person reading is inaccessible. What the reader is thinking is produced by the page, but the page is neither active nor interactive. If the reader arranges her time with care, she may be inaccessible by telephone or E-mail or any electronic medium. No sidebars on the page blink with reminders to call or purchase or click. The book is not wired to a consumer-research company studying the reading habits of people of her age and income bracket broken down by ethnicity, class origin, and self-definition.[1] The book may be borrowed and the author may be dead; the author may have never lived. The reader's inaccessibility threatens the structure of the social economy that possesses her in the twenty-first century. If she can manage to escape from the constant calls to attention, if she can maneuver out of her training to convince her that she has less time than previous generations, if she can strike through the ideological imperatives of her age that keep telling her how busy she is, then her own sense of now will expand steadily in the rich silence of the printed page held in her hands.

Among the scarcest resources in the twenty-first century is the present. The past looms behind us, richly documented by increasingly sophisticated historical methods and perspectives, steadily expanding as various interest groups tell their own histories and revise established historical paradigms by the lights of new ways of interpreting the past. The future is so vast that anxiety levels increase simply contemplating the horizons produced by technologies barely conceivable a decade ago. The turn of the calendar to the year 2000 seems to have produced another world of time, the psychological effect of all those zeros sustaining a sense of the immensity of what lies ahead. What is sandwiched between these equally expansive temporalities is the now, the middle time like the middle child, neglected and nearly forgotten in our efforts to know the past and discover the future. Contemporary information technologies act to attenuate the present by filling it with reminders and directions of what to do next, they thrive on deferral, and the human race may be heard in the heavens, chanting, "I'll get back to you," and "Yes, I got your message. Look for mine later today." We fill the now in preparation for tomorrow and in the completion of what was promised yesterday. In our multi-tasking environments, we simultaneously respond and chart and edit and search, keeping all those Windows open to allow the past and the future to enter freely and permeate the present.

Being busy has reached epidemic proportions in contemporary society. Preschool children report feeling stressed, without enough time in the day—even though they cannot yet tell time. They may not know what time it is, but they have succumbed completely to the ailment of nowlessness. Like their parents, they must continuously prepare, get ready; they are told by television and video programming aimed at their intelligence that they must watch this or do that or they will miss out. Their lives are guided by a vast, all-encompassing repair manual, directing thought, guiding action. Children mimic adults who have become thoroughly infected with the atemporality virus of nowlessness. They say they feel distracted, but it has long since been forgotten what it is they are distracted from, what they could possibly be doing instead of feeling pulled in multiple cognitive directions. There is so much to attend to now. This is the Information Age: corporations that manufacture information technologies tell us so. Minds today learn to play with information and become adept at gathering and experiencing that which is essentially inconsequential. Entertainment is likely to involve the pleasant experience of knowing something, observing something, processing information that requires no consequential choice save perhaps the decision to purchase on some deferred-payment plan.

The illusion of knowing steals time from the mind distracted by the seductions of infotainment. The content of information technologies is data driven and depends on what the mind can consume as input. Wherever one clicks on the screen, one is directed to some programmed site of consumption. The infinity of electronic consumption opportunities reverses the proportion of stimulus and response associated with reading. A book is a finite stimulus. It can be read repeatedly and interpreted innumerably, but as stimulus it remains static. The mind must respond to it creatively, must tinker with it and juxtapose its contents to other stimuli recalled or at hand. Its pieces are strewn across the mind like spare parts. While reading, the book is finite and the imagination infinite: there is quite literally no end to what can be thought in response to reading, especially reading imaginative literature. The proportion is reversed with electronic reading experiences. Here the screen provides infinite stimulation. The mind does not respond creatively but consumerly; its imaginative capacity is limited to discovering new links and new sites to know about. Sensitive to critics who warn against spending too much time at the screen or watching the clock for limited access hours, viewers may work against reverie as they try to cram as much clicking time into their experience.

Electronic information-gathering feels like knowing, feels like creative activity because it is marked by repeated discoveries of what was not known before. However, because the imagination is essentially shut down or limited to responsiveness, such experiences of knowing are limited entirely to the physical, material existence of what is. Nothing can be discovered that was not put there by someone else. The world of creation, the experience of now-enhancing reveries that depend on the mind's playful manipulation of its time and place, this dimension of cognition is shut off at the power switch that boots up the electronic world of information and boots out the imaginative capacity to escape it.

Ralph Waldo Emerson understood reading as the intersection of materialism and idealism, an act of suspension between the world of things and the world of thought where reading moves the mind from the dimension that contains consciousness to the dimension produced by the unlimited domain of the imaginative. In his essay "The Poet," Emerson equates the value of literary engagement with intellectual agitation. Inspiration, not conclusion, is the mark of a good read. "An imaginative book renders us much more service at first, by stimulating us through its tropes, than afterward when we arrive at the precise sense of the author" (258). The author's intent, long abused by twentieth-century literary criticism, carries a suspicion older than New Criticism. Authorial intent in formalist literary study is set aside not simply because it is often unknowable or even because it is irrelevant, though both notions have validity. Authorial intent is set aside because it threatens the idea of imaginative stimulation. To ask what the fiction "really means" is a query that is out of order in literary study. It is a proper question for the historian, the sociologist, and the biographer but not the literary critic or the serious reader of fiction and poetry. The value, or what Emerson termed the "service," of imaginative literature is intellectual stimulation rather than closure and not conclusion. "I think nothing is of any value in books excepting the transcendental and extraordinary" (Emerson 258): flights that project the mind into the parallel species of distracted intellectual existence, into realms of provisionality and contingencies of meaning. To become distracted by intellectual stimulation is to become unmoored from the harness of material precision and set loose on the extraordinary potential of the mind to imagine itself elsewhere—and to imagine the elsewhere here. Emerson's reader is a speculator, mining imaginative literature for what he knows will have value once articulated in the (reading) light of day.

The images that come to mind when the mind is garnished for good, useful reasons evoke conjunction and completion: I am held, held by this. I've got it; I grasp, hold, control; I digest it thoroughly and possess it. And mostly: I am doing what I am supposed to be doing. To the poet, the world is not conclusion, but the world, as it functions, thrives on its ability to create the illusion of completion. Read a little, do a little: fix the bicycle. Conversely, the images that arise when the mind strays evoke individuation and incompletion: I wander, I don't know where. I don't get it, I'm lost. As a result, the mind is no longer at the service of the real: it eludes me, it falls through my hands and so remains elsewhere, as do I. And so: I am doing something when I should be doing something else, and I am not getting anywhere. Communities root in conjunctive purposes, where everyone may not see eye to eye but at least they are looking at the same thing, jointly attentive, held fast to communal purposes. The education of the young, orderly market procedures, traffic patterns—all such systems are designed to assure that minds in common are tuned to sources of instruction and direction. Individuality, however, roots in disjunction, in privacy and in differentiation, where the isolate looks away from the community's purpose, fingering the fire poker and contemplating the absent whale line, reading in the age of distraction. Minds are justifiably pulled in two directions whenever they are pulled at all: toward what social obligations require and toward distraction. Or, alternatively, they are pulled toward pleasurable patterns of cognition and toward social distraction, which nags. *Distraction,* in its most common usage, is defined as what the mind is doing when it has other, pressing concerns. Enjoying the pleasures of the breakfast table, the mind is distracted by the clock and *it's time to go to work.* At work on an important project the mind is distracted by the memory and *it starts to think about home.* No one's intellect is garnished for long, and all minds travel in a cycle of being held fast and released, held fast by the release and released again.

Readers read to be held and to escape, at once engaged and suspended. Here literary criticism has not served readers well in recent decades. Concerned, rightly so, with rescuing the idea of the literary in a culture that grants tremendous privileges to empiricism, much literary criticism has been consumed with public, socially justifiable action. It has offered a succession of pre-textual models and interpretive performances in a program that seeks generally to counter or at least stand equivalent to scientific truth claims. Fresh interpretations of texts may bring pleasure by inspiring new readings, but holistic readings may risk destroying the object of interpretation by foreclosing on subsequent encounters. Interpretation is as inevitable as the exclamation "What a fine sauce!" to explain what made the meal so enjoyable. But a listing of ingredients may not be a very useful adjunct to the sensual and very private experience that inspired the articulation of pleasure. Similarly, literary criticism that stops with interpretation will not provide the rationale for a continuous program of reading, especially in an age where distractions reign and the dominant consumer ideology encourages the conclusion that there is no time to read. Instead, literary analysis has an obligation to articulate the status of the imaginative in written form as a highly particularized, private alternative to what drives mass society and fuels its inevitable efforts to make experience quantifiable.

In addition to interpretation, then, let us make room for textual speculation and the studied expansion of the now. Let us reevaluate the public use of private experience, the Emersonian dictum that "nothing is of any value in books excepting the transcendental and extraordinary." The necessity of such endeavor is particularly apparent in educational institutions, where, in theory if not in practice, the potentiality of interior, imaginative energies are cultivated, nurtured, and encouraged. Imaginative speculation is the province of the arts and the humanities, where how to look, how to read, and how to listen constitute the survival skills of serious programs of willful distraction. Interpretation guided by the speculative overturns the traditional value of reading, which stresses comprehension, with sustained consideration of what happens to the imaginary during the experience of fiction, poetry, and the arts generally. "What have you read lately?" replaced by "What stimulates these days?" in the chatter of the cocktail hour. The question is one of stress: What is the value of reading imaginative works? Preserving expansive, speculative space demands vigilance in structures of mind trained more readily for the lockstep usefulness of reception. The world as found has a great interest in maintaining itself; resistance to entrenched systems is cultivated by the will of interior reserve.

Why limit the mind to what *is* and, worse, to what is available as information in the public domain? One's experience of the now is likely to be thoroughly colonized by consumer and production predators whose goods and services are made sexy by their novelty and by their claims to meet pressing if newly discovered needs. The introduction of car phones models this process. Originally, the automobile, in addition to its status as a revolutionary instrument of travel, carried symbolic implications of freedom and escape from obligation. Teenagers since the 1950s have used the car to find the solitude so necessary to adolescence. To adults en route between work and home, the car was a refuge, an untouchable duration between familial and employment obligations. Even shuttling kids around offered an escape from the drudgery of housework, a taste of freedom to the housewife in the family's second car. The telephone, conversely, is a communication technology, fostering one-to-one exchange, revolutionizing communication in social and business situations by introducing information as a shared commodity. On the phone, one assumes a reasonable percentage if not the undivided attention of one's listener. The telephone symbolizes the potential of engagement with others; in the household, it has represented (until very recently) the single interactive link to the social world.

In the late twentieth century, marketers convinced consumers that it was necessary to have a phone in their cars and to accept the redefinition of their automobiles. The car was transformed from solitary, private space to the site of absence—a place where one was out of touch and in danger of missing something. The automobile is no longer a space of freedom but a transportation device that is considered dangerously incommunicado without its cell phone. In brief, the car has been colonized. When equipped with a phone, it is no longer on the road but in the network—fully wired and connected to information sources that demand responses. Consumer-marketing images of stranded women, lost teenagers, and delayed fathers easily convinced drivers of the need for such technology. The constant in information technology is that it contracts the individual's experience of now by checking the spectrum of unlimited individual speculation with the idea of absence. Communication technologies introduce into attention spans the notion that everything, everywhere, is absent and so in danger of being missed unless attended to. One is encouraged thus to consider unmediated solitude less an opportunity for speculative flight and more the condition of being temporarily out of touch until the mind can again be plugged into sources of information that will eliminate immediacy. The car has slipped into this paradigm, now defined as the site of absence, not escape, where the driver finds herself wholly out of touch, stranded, and alone. The phone sits nestled between the seats and calls no one in particular, but it calls us all to attention, contracting the expansive now of the commuter's solitude, linking the driver to a network of exchange potential.

Cultural critics have thus observed that much of the silence of life, the realm of personal reflection and spirituality, has been colonized in the twentieth century. The automobile is only one example. Mark Slouka invokes a natural metaphor: "Erosion and encroachment, forces generally associated with natural world, with the slow-motion wearing of rock and earth, or the burial of a meadow under asphalt or scrub, apply to the supernatural one as well. At the close of the twentieth century, the terrain of the spirit—by which I mean the domain of silence, of solitude, of unmediated contemplation—is everywhere under siege, threatened by a polymorphous flood of verbal and visual signals (electronically generated, predominantly corporate), that together comprise what we might call the culture of distraction" (148). A nineteenth-century individual, living as late as the 1920s, knew silence as the context of thought, conversation, and general existence. Noises and voices were associated with and usually connected to their physical sources. A book spoke to the mind in language of some volume in such circumstances, the authorial voice not so much a distraction but a way to fill silence with stimulation, inform thought with content, simulacra, and cognitive patterns. Under such circumstances, speculation found the room it needed to expand the present with its sense of endless possibilities.

The voice of imaginative literature is far less boisterous today, in the context of mass-media blare, and instead of filling silence, the act of reading produces silence, distracting the mind away from the incessant input of mediated sources and constructing a postelectronic void. The strain on the will is much greater as imaginative energies must be expended to create the circumstances necessary to speculative endeavors. Distraction and attention are reversed: media sources in their infancy distracted us from silence—the phone rang, the records played, the radio got turned on for a while. Now that media sources are ubiquitous, silence distracts us from the constancy of electronics. Some people cannot tolerate the quiet when it occurs. Should one read with background noise? To comprehend in such settings, one must be able to enforce an interior silence. Only then, and only if the mind can sustain its inattention to the blare, does that inner silence, that interior void, become influenced by another mind's creativity. How many of us have read pages only to realize that the reading had produced a void but had not filled that void with content, and so we have no idea what we have read. We cleared the field, but the seeds blew away, taking our attentiveness with it.

Reading in the Age of Distraction is strenuous business. The television holds us fast, moors us to its insistence that we stay tuned, get the latest information, and not miss the next, even more vital, installment. A middle-class American household is a communication node on a predator's web, where telephone and postal services, cable and Internet access form an alliance to electronically subvert the Fourth Amendment's promise of safety and security at home. The police may need a warrant, but no one else does. How does one take imaginative flight under such conditions? The personal computer screen continuously beckons us to click away from cognitive pressure, so if our thoughts stall, the mouse brings another mediated realm before us. What feels like interaction is more accurately considered a kind of enforced passivity, ensuring that the mind will not be overtaxed. To read speculatively in this environment is to place oneself at risk, to become distracted from the web of connection to what is happening now, as if to ratchet oneself effectively out of the loop of attention constructed by the limitless potential of electricity and speed. To be moored assures the security of knowing where you are and what you should be thinking and doing. Pay attention! But to be unmoored (to be distracted by what is outside the window while the speaker presents his findings or to stare silently into the book while the media blare and beckon)—to be unmoored is to enter another realm of existence, eternally unfinished, inconclusive, and unsatisfied, the home of desire. The clocks tick, the beepers beep, the pagers page and we know, all the time, what we should be doing unless . . . unless we become distracted.

The will to distraction can be nurtured only by saying no to something that threatens more people than drug addiction or whatever it was we were told to "just say no" to at one time. Educators in the humanities who think they are helping their students to learn by turning classrooms into media centers contribute to the erosion of thinking by linking individual thought to mediation. When students who cannot tolerate silence are drugged (by prescription, that is), their minds become satisfied with less stimulation; they are in training for docility. Households where televisions contribute to a constancy of background noise are steadily abusing the intellects cultivated there by suppressing contemplation, making it not simply impossible but uncomfortable to think. Under such conditions, passive receptivity comes to feel like thought, and silence, if it functions at all, functions as a pained distraction. Speculative thinking seems especially dangerous to minds seeking to be moored. Nonetheless, it is impossible to eliminate the technological environment, and it would make no sense to try. I am no Luddite, as I write on my word processor, with my E-mail account and the household televisions sets with their VCR spouses poised for attentive reception. It is not a question of returning to another era or re-creating an anachronistic relationship between silence and noise. On the contrary, the benefits of technology are too valuable to argue against or resist. Nonetheless, when campers go into the wilderness to enjoy the wild, they do not go without equipment and training. Similarly, when we enter the technological, information environment, we are foolish to do so without being comparably equipped for that existence. Humanists have no business holding back the tide of technological progress and inevitability. The duty of humanists since Socrates is not the enforcement, creation, or maintenance of the conditions of existence. Instead, our concern is how to live well within these conditions and where to locate sanctuaries from them. In the Age of Distraction, how does one possess and cultivate an imaginative and speculative intellectual life worth living?

We need to rethink distraction. The direction taken by the mind, the way the mind attends, is properly the concern of those whose industries—educational, governmental, entertainment—are directly implicated in the formation and direction of distracted minds. Those in the literary professions must be particularly attentive. Imaginative literature evokes and exists within nonmateriality. It articulates ideas, images, and emotional reactions tied solely to simulacrum rather than to the mediated compulsions of "real life." Literature demonstrates page after page that there is more, there must be more, and thus suggests and speculates on the existence of alternatives to whatever predicament in which we find ourselves. At the same time, there is nowhere to click to beyond one's own imaginative resources to make sense of the written word. Literary experience transforms habits of cognitive distraction into art forms by seizing on the mind's desire to be elsewhere and to be released from physical limitation. The mind needs distraction in the way the body needs exercise: not only to survive, but to extend its life and make itself durable, renewable, agile. Reading imaginative literature is in itself speculative thought. The mind engages in the contemplation of what does not exist but what, by virtue of being thinkable, does in some sense exist as a species apart. Such hypothetical reflection expands the possible, and as the possible expands so do the conditions of our survival as a species on this fragile planet.

The task of literary criticism includes making public the value of literary study, arguing and demonstrating the rationale for resisting the blare and allowing the distractions of imaginative literature a place within the swirling demands on contemporary cognition. The twentieth century saw literary study wed increasingly to empiricism. Perhaps the twenty-first will see the study of imaginative literature linked to the arts of conjecture, meditation, and the capacity to surmise. We might think of literary critics—from public-library reading groups to MLA membership—as what Charles Altieri called "the community that becomes available among those who can specify what they admire or find useful in that complex authorial labor." Reading imaginative literature is a "process of engaging another's articulation of emotions" that "allows us to see how deeply our own structures for intense, passionate life are woven into those figures" ("Values" 83). For reading, the mind's desire for individuation is subsidized by the book's articulation and its invitation to tinker with all of its imaginative spare parts. In the same way that a nascent mind develops through intense interaction with its caretakers and educators, the intellect is renewed by close connection to sustained, imaginative intensity.

THE SOUND THE READER SEES

Isabel finally confessed that it wasn't like living with a person at all, it was like living with a sound. And the sound didn't make any sense to her, didn't make any sense to any of them—naturally. They began, in a way, to be afflicted by this presence that was living in their home. It was as though Sonny were some sort of god, or monster. He moved in an atmosphere which wasn't theirs at all. They fed him and he ate, he washed himself, he walked in and out of their door; he certainly wasn't nasty or unpleasant or rude, Sonny isn't any of those things; but it was as though he were all wrapped up in some cloud, some fire, some vision all his own; and there wasn't any way to reach him
James Baldwin

A sound is an aural image, evocative but nonmaterial, the raw stuff of the composer's vision and the musician's craft. James Baldwin casts Sonny more like an aesthetic image than a person, more like a literary character than some person. And like a literary character, "there wasn't any way to reach him" save by contemplation, by the act of speculating *human*. Nonetheless, Sonny, sharing the quality of sound, distracts. His presence afflicts, affects, and commands attention while his hearers, perhaps, ought to be attending something else but choose against resisting the meditation Sonny inspires. The only "fact" in this passage is the reader, his mind and cognitive capacities, the sounds in the mind developed before the encounter with Sonny and evoked by the language that conjures him within specific patterns of thought. Like the reader, Sonny wants out and he wants in, he attends and he is distracted, he knows and he speculates.

The interior life of every human being is characterized by an intensely felt resistance to the conditions of its external existence; it cannot be sustained on information, on preparation, or by the group life of the household, the corporation, or the team. Desire is the sound the reader's mind encounters and produces as it speculates on the real life of the imagination, wrapped up in some cloud, some vision of how it is, alternatively, elsewhere. The species beyond has been called the realm of the spirit, the soul, the imaginary, the subconscious—the mind has speculated continuously about what distracts it from the world inhabited by its body and by its mortality. The sustaining capacity for speculation is too valuable to a life of integrity to leave it to commercial exploitation. The humanities alone are equipped in methodology and resources to provide the kind of practice and training to ensure a lifetime of speculative, imaginative, responsive exploration; to establish the now of reflective meditation and to stake its claim to cognitive territory, secure from the encroachments of digitized colonization. Foremost in the mission of the humanities in the twenty-first century is to declare independence from the empirical world, to locate in its irrelevance to what is established the vitality of its purpose and its value to the maintenance of human beings' autonomous existence.

It is a truism among students who study literature today that everyone who reads a novel or a short story receives a distinctive, equally valid meaning from the experience. The radical relativism of the position is comfortable for an undergraduate in a literature class because it means that anything goes, and there can be no wrong answer or invalid interpretive reading. The concept and the fear of the wrong answer is imported from social science, math, natural science, and professional-training classes, where the concept of being wrong is applicable and the fear reasonable. Of course, it is possible to get it wrong in a literature class, but even so, the student is less likely to be asked to recall specific, verifiable facts than to venture with more credibility into the realm of interpretation. Once in this realm, the student may get the idea that anything goes, especially if the instructor brings meanings to the text that the student would never have dreamed of on her own. However, the student defense of radical relativism is based on a misunderstanding of literary study. It is not the case that there are no right or wrong answers, but it is the case that the binary thought processes enforced by the concept of right and wrong are antithetical to literary experiences.

Concepts of true and false, fiction and nonfiction, right and wrong, are strangers to literary employments, as they are to all speculative forms of intellectual endeavor. Altieri suggests that literary studies "provide vital alternatives to the perspectives on value established by empiricist and utilitarian conceptual frameworks" ("Values" 67). As the Age of Distraction brings reproducible and predictable experience into the masses' lives, alternative modes of consciousness may indeed become endangered phenomena, leading some critics to summon back the "ghost of the unmediated world" (Slouka 153) for relief from commodified thought. The increasingly defensive posture taken by those who defend literary endeavors would indicate that while tolerance of alternatives declines, the need for them, paradoxically (but also, in a way, logically), rises. Gary Morson invokes the idea of "sideshadowing" to get at this notion. Foreshadowing is a well-known feature of classroom literary instruction, referring to forthcoming events or occurrences in the narrative that can be verified (and their detection judged right or wrong). But sideshadowing is proffered to evince the value of all literary endeavor. "Sideshadowing restores the *possibility of possibility*. Its most fundamental lesson is: to understand a moment is to grasp not only what did happen but also what might have happened. Hypothetical histories shadow actual ones. Some nonactual events enjoy their own kind of reality: the temporal world consists not just of actualities and impossibilities but also of real though unactualized possibilities. Sideshadowing invites us into this peculiar *middle realm*" (119). Thus, reading imaginative literature habituates the mind to a constant and reflexive consideration of alternatives. Morson's idea of sideshadowing makes emblematic not relativism but a kind of intellectual or dimensional speculation. Readers know that some characterizations, situations, and images from what they have read are more real, more vital, than many of the actualities of their lives. There are also recurrent potentialities, the *might have beens* in literary representation, that exist solely as speculation. At the very least, literary encounters, in Morson's words, "cultivate the imagination of alternatives" (273), including, not least of all, a sustained parallel intellectual existence, signaled and memorialized by the books we've read. These encounters represent entry to the species beyond the world that claims conclusion.

Without a sense of alternatives, existence becomes intolerable. Each of us requires, for intellectual sustenance, a speculative existence apart from those closest to us, including those from whom we feel inseparable. In an essay on Katherine Mansfield, Willa Cather refers to the double life of family members in even the most harmonious of homes. There is the "group life, which is the one we can observe in our neighbor's household," the one that may be filmed and commodified, we might add, for national consumption. Parallel to the life of the group lies "another—secret and passionate and intense—which is the real life that stamps the faces and gives character to the voices of our friends. Always in his mind each member of these social units is escaping, running away, trying to break the net which circumstances and his own affections have woven about him" (*Not under Forty* 136). The group life may be augmented and perhaps even improved by visible things: a bigger house, a backyard, a home entertainment center. But the "real life" of the household must be nurtured by constructions made of finer, more esoteric, transcendent materials. Escape is vital, and exit routes into speculative intellectual existence must be marked clearly, or those minds will deflate into mediocre lives of petty distraction and the recklessness of desperation.

Surviving intellectually in the Age of Distraction is an act of diligence, of resistance to intellectual commodification, to consumer predators, and to being counted solely as an interest within a mass audience identified on the crosshairs of some grid of potential use. When we teach distraction, we encourage taking control of our attention spans and confronting the central mystery of what commands our attention. When we consider the history of distraction, we recognize it as a trope for a specific kind of freedom: the freedom to tinker with imaginative resources, sometimes with good reason, sometimes for the hell of it. A repair manual guides toward specific ends; a spare part is something to tinker with. Feminists have asked women why they do not own their bodies, why they surrender their bodies to the state and to the men who name them. Marxists ask workers why they do not own the means of production. Psychologists urge ownership of our neuroses. A distractionist asks why men and women do not own their attention, why they surrender their minds to externally defined and timed distractions, to the media interests that broadcast and transmit images that comprise and direct. Verily, the further from grounded reality the mind can fly, the more powerful it will become when it is time (and there is always time) to return to the demands of material existence, to the realities as we know them in our present time. Imaginative writing can thus provide wings of distraction to minds riveted by mediation. While the present is tabulated like a bill of goods for quick consumption, the mind may continue to cultivate sites of resistance and, within the interiors of its own thought patterns, negotiate the terms by which its imaginative life shall endure its time on earth.

NOTES

INFORMATION, PLEASE: THE DISTRACTIONS OF THE DIGITAL ENVIRONMENT

1. Educational issues are taken up in greater detail in "Teaching to Distraction," where this argument is presented more completely.

THE PUBLIC VALUE OF DISTRACTION

1. I do not mean to disparage the many good teachers of freshman composition, many of whom grapple with these central contradictions.

2. Among the examples: Stephen Hahn and Arthur F. Kinney, eds., *Approaches to Teaching Faulkner's* The Sound and the Fury (New York: Modern Language Association of America, 1996); Stephen M. Ross and Noel Polk, eds., *Reading Faulkner:* The Sound and the Fury (Jackson: University Press of Mississippi, 1996); Noel Polk, ed., *New Essays on* The Sound and the Fury (New York: Cambridge University Press, 1993); John T. Matthews, The Sound and the Fury: *Faulkner and the Lost Cause* (Boston: Twayne, 1991); Harold Bloom, ed. *William Faulkner's* The Sound and the Fury (New York: Chelsea House, 1988); Arthur F. Kinney, ed., *Critical Essays on William Faulkner: The Compson Family* (Boston: G. K. Hall, 1982); Andre Bleikasten, *The Most Splendid Failure: Faulkner's* The Sound and the Fury (Bloomington: Indiana University Press, 1976); James B. Meriwether, comp., *The Merrill Studies in* The Sound and the Fury. (Columbus, OH: Merrill, 1970); Michael H. Cowan, ed., *Twentieth Century Interpretations of* The Sound and the Fury: *A Collection of Critical Essays* (Englewood Cliffs, NJ: Prentice-Hall, 1968).

3. This has become a problem in the workplace. See for example, Breuer. It is also an issue with the growing move to allow employees to work at home. See Roha.

SPECULATION AND SURVIVAL IN THE AGE OF DISTRACTION

1. I refer to traditional paperback and cloth books, not to electronic book technologies, which are designed to perform the functions enumerated and more.

WORKS CITED

Adams, Henry. *The Education of Henry Adams.* 1918. Ed., intro. and notes Ernest Samuels. Boston: Houghton Mifflin, 1973.

Altieri, Charles. *Canons and Consequences: Reflections on the Ethical Force of Imaginative Ideals.* Evanston: Northwestern University Press, 1990.

——"The Values of Articulation: Aesthetics after the Aesthetic Ideology." *Beyond Representation: Philosophy and the Poetic Imagination.* Ed. Richard Eldridge. Cambridge Studies in Philosophy and the Arts. New York: Cambridge University Press, 1996. 66–89.

Altman, Rick. "Television/Sound." *Studies in Entertainment: Critical Approaches to Mass Culture.* Ed. Tania Modleski. Theories of Contemporary Culture. Bloomington: Indiana University Press, 1986. 39–54.

Apple, Michael W. "Cultural Capital and Official Knowledge." *Higher Education under Fire: Politics, Economics, and the Crisis of the Humanities.* Ed. Michael Bérubé and Cary Nelson. New York: Routledge, 1995. 91–107.

Aronowitz, Stanley. *Roll over Beethoven: The Return of Cultural Strife.* Hanover, NH: University Press of New England, 1993.

Baldwin, James. *Going to Meet the Man.* 1965. New York: Random House, 1993.

Bérubé, Michael. *Public Access: Literary Theory and American Cultural Politics.* New York: Verso, 1994.

Birkerts, Sven. *The Gutenberg Elegies: The Fate of Reading in an Electronic Age.* New York: Fawcett Columbine, 1994.

Boorstin, Daniel, ed. *An American Primer.* New York: Penguin, 1984.

Breuer, Nancy. "Minimize Distractions for Maximum Output." *Personnel Journal* May 1995: 70–74.

Bridges, William. *JobShift: How to Prosper in a Workplace without Jobs.* Reading, MA: Addison-Wesley, 1994.

Burstein, Daniel, and David Kline. *Road Warriors: Dreams and Nightmares along the Information Highway.* New York: Dutton/Penguin, 1995.

Carafiol, Peter. *The American Ideal: Literary History as a Worldly Activity.* New York: Oxford University Press, 1991.

Cather, Willa. *My Ántonia.* 1918. New York: Penguin, 1994.

————*Not under Forty.* 1922. Lincoln: University of Nebraska Press, 1988.

————*Obscure Destinies.* New York: Vintage Books, 1974.

————*On Writing: Critical Studies on Writing on Art.* 1920. Lincoln: University of Nebraska Press, 1920.

Chance, M. R. A., and R. R. Larsen. "Introduction." *The Social Structure of Attention.* Ed. Michael R. A. Chance and Ray R. Larsen. New York: Wiley, 1976. 1–10.

Cherkes-Julkowski, Miriam, Susan Sharp, and Jonathan Stolzenberg. *Rethinking Attention Deficit Disorders.* Cambridge, MA: Brookline, 1997.

Clinton, William J. "Administration of William J. Clinton, Inaugural Address of January 20, 1993." *Weekly Compilation of Presidential Documents* 29 (January 25, 1993): 75–77.

Crèvecoeur, J. Hector St. John de. *Letters of an American Farmer and Sketches of Eighteenth-Century America.* Ed. Albert E. Stone. New York: Penguin, 1981.

Derrida, Jacques. "Declarations of Independence." *New Political Science* 15 (1986): 7–15.

Dickinson, Emily. *Final Harvest.* Selected and intro. Thomas H. Johnson. Boston: Little, Brown, 1964.

Donovan, Josephine. "The Pattern of Birds and Beasts: Willa Cather and Women's Art." *Writing the Woman Artist: Poetics, Politics, and Portraiture.* Ed. Suzanne W. Jones. Philadelphia: University of Pennsylvania Press, 1991. 81–95.

Eidelberg, Paul. *On the Silence of the Declaration of Independence.* Amherst: University of Massachusetts Press, 1976.

Emerson, Ralph Waldo. *The Portable Emerson.* Ed. Carl Bode. New York: Penguin, 1981.

Faulkner, William. *Essays, Speeches, and Public Letters.* Ed. James B. Meriwether. New York: Random House, 1965.

————*Go Down, Moses.* 1942. New York: Vintage, 1990.

————*Light in August.* 1932. New York: Vintage, 1990.

————*Sanctuary.* 1931. New York: Vintage, 1993.

————*The Sound and the Fury.* 1929. New York: Vintage, 1990.

Fetterly, Judith. "*My Ántonia:* Jim Burden and the Dilemma of the Lesbian Writer." *Gender Studies: New Directions in Feminist Criticism.* Ed. Judith Spector. Bowling Green, OH: Bowling Green State University Popular Press, 1986. 43–59.

Fliegelman, Jay. *Declaring Independence: Jefferson, Natural Language, and the Culture of Performance.* Stanford: Stanford University Press, 1993.

Fossum, Robert H., and John K. Roth, eds. *An American Primer.* New York: Paragon, 1988.

Free [Abbie Hoffman]. *Revolution for the Hell of It.* New York: Dial, 1968.

Gerber, Scott Douglas. *To Secure These Rights: The Declaration of Independence and Constitutional Interpretation.* New York: New York University Press, 1995.

Giroux, Henry A. *Channel Surfing: Race Talk and the Destruction of Today's Youth.* New York: St. Martin's, 1997.

Guillory, John. *Cultural Capital: The Problem of Literary Canon Formation.* Chicago: University of Chicago Press, 1993.

Hans, James S. *The Mysteries of Attention.* Albany: State University of New York Press, 1993.

Hobsbawm, E. J. *Nations and Nationalism since 1780: Programme, Myth, Reality.* New York: Cambridge University Press, 1990.

Hooks, Bell. *Teaching to Transgress: Education and the Practice of Freedom.* New York: Routledge, 1994.

Huxley, Aldous. *Brave New World Revisited.* New York: Harper and Row, 1958.

Illich, Ivan. *Toward a History of Needs.* New York: Pantheon, 1978.

Iser, Wolfgang. *The Fictive and the Imaginary: Charting Literary Anthropology.* Baltimore: Johns Hopkins University Press, 1993.

Jay, Gregory S. *American Literature and the Culture Wars.* Ithaca: Cornell University Press, 1997.

Jayne, Allen. *Jefferson's Declaration of Independence: Origins, Philosophy, Theology.* Lexington: University Press of Kentucky, 1998.

Jewett, Sarah Orne. *The Country of the Pointed Firs and Other Stories.* New York: Penguin, 1995.

Koch, Adrienne, and William Peden, eds. *The Life and Selected Writings of Thomas Jefferson.* New York: Modern Library, 1944.

Krakauer, Jon. *Into the Wild.* New York, Doubleday, 1996.

Lambert, Deborah. "The Defeat of a Hero: Autonomy and Sexuality in *My Ántonia.*" *American Literature* 53 (1982): 676–90.

Levine, Lawrence. *The Opening of the American Mind: Canons, Culture, and History.* Boston: Beacon, 1996.

Lipset, Seymour Martin. *American Exceptionalism: A Double-Edged Sword.* New York: Norton, 1996.

London, Jack. *The Call of the Wild.* 1903. New York: Bantam Classic, 1981.

Looby, Christopher. *Voicing America: Language, Literary Form, and the Origins of the United States.* Chicago: University of Chicago Press, 1996.

Maier, Pauline. *American Scripture: Making the Declaration of Independence.* New York: Knopf, 1997.

Manguel, Alberto. *A History of Reading.* New York: Penguin, 1996.

Marc, David. *Bonfire of the Humanities: Television, Subliteracy, and Long-Term Memory Loss.* The Television Series. Ed. Robert Thompson. Syracuse: Syracuse University Press, 1995.

Martin, Terence. *Parables of Possibility: The American Need for Beginnings.* New York: Columbia University Press, 1995.

McKibbon, Bill. *The Age of Missing Information.* New York: Penguin, 1993.

McLaughlin, Thomas. *Street Smarts and Critical Theory: Listening to the Vernacular.* Madison: University of Wisconsin Press, 1996.

Melville, Herman. *Moby-Dick; or, The Whale.* 1851. New York: Penguin, 1992.

Minnich, Elizabeth Kamarck. "'What is at Stake': Risking the Pleasures of Politics in a Democratic Education." *Soundings* 80.2.3 (summer/fall 1997): 251–64.

Morson, Gary Saul. *Narrative and Freedom: The Shadows of Time.* New Haven: Yale University Press, 1994.

Mukherjee, Bharati. *Jasmine.* New York: Ballantine, 1989.

Nagel, Thomas. *The View from Nowhere.* New York: Oxford University Press, 1986.

New Shorter Oxford English Dictionary on CD-ROM. Oxford: Oxford University Press, 1996.

Nussbaum, Martha C. *Cultivating Humanity: A Classical Defense of Reform in Liberal Education.* Cambridge: Harvard University Press, 1997.

———*Poetic Justice: The Literary Imagination and Public Life.* Boston: Beacon, 1995.

Parker, David. *Ethics, Theory, and the Novel.* New York: Cambridge University Press, 1994.

Postman, Neil. *Amusing Ourselves to Death: Public Discourse in the Age of Show Business.* New York: Viking, 1984.

Rawls, John. *Political Liberalism.* New York: Columbia University Press, 1993.

Reed, Edward S. *The Necessity of Experience.* New Haven: Yale University Press, 1996.

Roha, Ronaleen. "Home Alone." *Kiplinger's Personal Finance Magazine* May 1997: 85–90.

Slouka, Mark. "In Praise of Silence and Slow Time: Nature and Mind in a Derivative Age." *Tolstoy's Dictaphone: Technology and the Muse.* Ed. Sven Birkerts. Saint Paul, MN: Graywolf Press, 1996. 147–56

Starker, Steven. *Evil Influence: Crusades against the Mass Media.* New Brunswick, NJ: Transaction, 1989.

Tichi, Cecelia. *Electronic Hearth: Creating an American Television Culture.* New York: Oxford University Press, 1991.

Tierney, Thomas F. *The Value of Convenience: A Genealogy of Technical Culture.* SUNY Series in Science, Technology, and Society. Ed. Sal Restivo. Albany: State University of New York Press, 1993.

Toll, Robert C. *The Entertainment Machine: American Show Business in the Twentieth Century.* New York: Oxford University Press, 1982.

Twitchell, James B. *Carnival Culture: The Trashing of Taste in America.* New York: Columbia University Press, 1992.

United States. President. *Inaugural Addresses of the Presidents of the United States from George Washington 1789 to Richard Milhous Nixon 1969.* 91st Cong., 1st sess., House Doc. 91–142. Washington: GPO, 1969.

Vaughan, Alden T., and Edward W. Clark. *Puritans among the Indians: Account of Captivity and Redemption, 1676–1729.* Cambridge: Belknap–Harvard University Press, 1981.

Virilio, Paul. *Speed and Politics: An Essay on Dromology.* Translated by Mark Polizzotti. New York: Semiotext(e), 1986.

———*The Vision Machine.* Bloomington: Indiana University Press, 1994.

Volpe, Edmond. *A Reader's Guide to William Faulkner.* New York: Farrar, Straus, and Giroux, 1964.

Walton, Kendell L. *Mimesis as Make-Believe: On the Foundations of the Representational Arts.* Cambridge: Harvard University Press, 1990.

Williams, Raymond. *Television: Technology and Cultural Form.* New York: Schocken, 1975.

Withington, Ann Fairfax. *Toward a More Perfect Union: Virtue and the Formation of American Republics.* New York: Oxford University Press, 1991.

Yippie Book Collective. *Blacklisted News, Secret Histories . . . From Chicago, '68 to 1984.* New York: Bleeker, 1983.